Mathematics and Computing/Technology
A Third-level Course
MT365 Graphs, Networks and Design

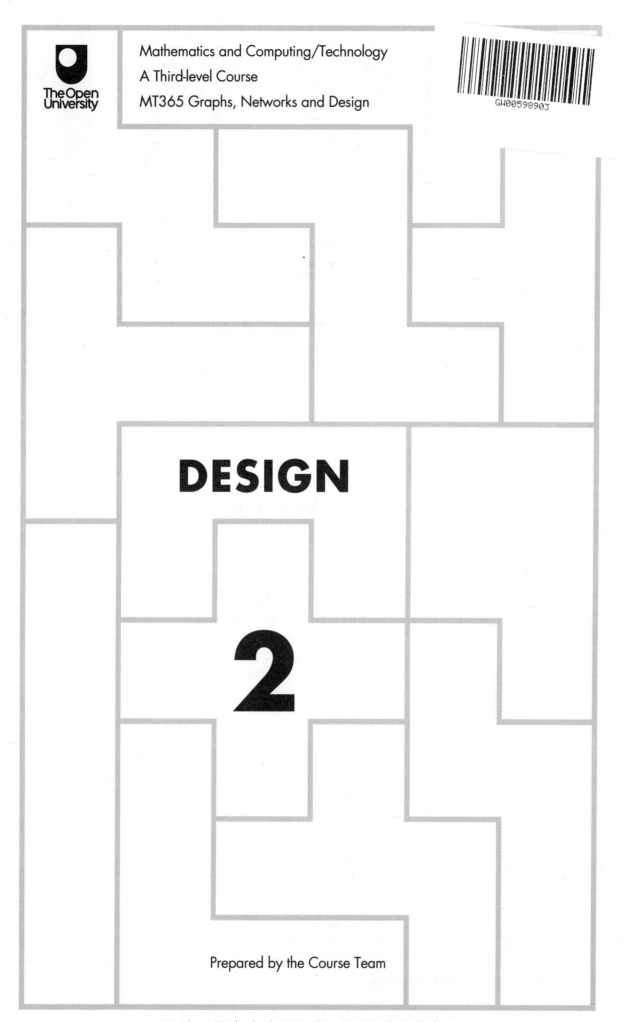

DESIGN

2

Prepared by the Course Team

KINEMATIC DESIGN

Study guide

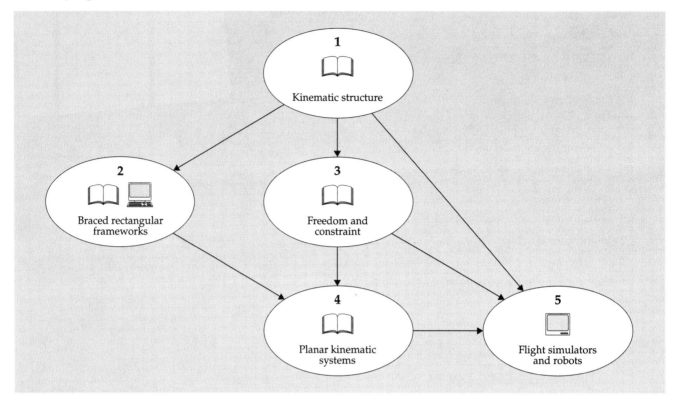

In Section 1 we introduce the basic concepts needed for an understanding of kinematic structure. Study this section first, and ensure that you become familiar with the ideas and terminology. Pay particular attention to the graph representations: the *direct graph* and the *interchange graph*.

In Section 2 we analyse in detail a special type of kinematic system that you met in the *Introduction* unit — the *braced rectangular framework*. The section may be studied in isolation, but is best tackled in the context of the ideas of Section 1. There are computer activities associated with Section 2.

Sections 3 and 4 are the most important sections. In Section 3 we examine the notions of *freedom* and *constraint*, which provide us with quantitative measures of the general kinematic behaviour of a system. In Section 4, we use these measures to derive several *mobility criteria*, which we then apply to kinematic systems.

Section 5 is a television section dealing with two practical kinematic systems — a robot manipulator and a flight simulator. The programme can be watched at any time, but is best studied after you have worked through Sections 3 and 4.

The Open University, Walton Hall, Milton Keynes, MK7 6AA.

First published 1995. Reprinted 1997, 1999, 2002, 2003.

Designed by the Graphic Design Group of the Open University.

Typeset in the United Kingdom by the Open University.

Printed in Malta by Interprint Limited.

ISBN 0 7492 2217 4

This text forms part of an Open University third-level course. If you would like a copy of *Studying with the Open University*, please write to the Central Enquiry Service, PO Box 200, The Open University, Walton Hall, Milton Keynes, MK7 6YZ. If you have not enrolled on the course and would like to buy this or other Open University material, please write to Open University, Educational Enterprises Ltd, 12 Cofferidge Close, Stony Stratford, Milton Keynes, MK11 1BY, United Kingdom.

MT365Design2i1.5

Contents

Introduction

This is the second of the Design units. It is concerned with a particular class of physical object — the so-called *kinematic systems*. Common examples of kinematic systems are mechanical diggers and loaders seen at roadworks, but, as we shall see later, there are many more familiar examples in the home and office. The primary function of such a system is to produce, transmit, control or constrain movement of one sort or another, and the class includes as a special case systems designed to resist movement (that is, *structures*). Common examples of structures are electricity pylons and roof trusses, but again we shall meet more familiar examples later.

In general, an investigation of the behaviour and construction of kinematic systems involves a thorough understanding of the detailed geometry, kinematics and dynamics of each system. These aspects require an extensive knowledge of the material properties of the components, and the correct application of all the relevant physical laws of motion.

Geometry is concerned with size and shape; kinematics with motion; dynamics with forces.

Here we are not primarily interested in the motion itself, nor in the means by which it is generated, but rather in the types of arrangement which give rise to various degrees of movement in a particular system. Furthermore, we treat the component parts of the system as ideal rigid bodies, so that we exclude elastic, plastic and fluid components from the outset, and do not explicitly consider the physical properties of the materials involved.

Our principal aim is to investigate ways in which various rigid mechanical components can be interconnected so as to form specific types of kinematic system. The main discussion centres on a description and classification of *kinematic structure*, by which we mean the arrangement, juxtaposition, and nature of interconnection of the component parts of a system.

We model kinematic structure in terms of various graph representations. We also give particular emphasis to the combinatorial aspects of the subject, in which the significant parameters are the number of components, the number of connections (or joints), and the degree of movement. The precise geometric details of shape, size, position and orientation of the components and joints are usually of secondary importance in this context, although this so-called *kinematic geometry* can introduce many complications. We discuss the significance of some of these.

In Section 1, *Kinematic structure,* we deal with the fundamentals of kinematic structure. We introduce and examine various kinematic systems to determine the primitive kinematic elements — *links* and *joints* — from which they are constructed. We then introduce two types of graph representation used to model kinematic systems — the *direct graph* and the *interchange graph*. Finally, we discuss the basic kinematic geometry which characterizes kinematic systems and their motions.

Section 2, *Braced rectangular frameworks*, deals with the problem of bracing rectangular framework structures in order to make them *rigid* — that is, so that they resist any movement. One solution to the problem involves the use of a *bipartite graph* to represent the structure. We shall see how *spanning trees* in the bipartite graph correspond to minimum bracings, and hence enable us to use bracing components most economically.

You met braced rectangular frameworks and their bipartite graphs in the *Introduction* unit.

In Section 3, *Freedom and constraint*, we deal with the concept of mobility. The *mobility* of a system is an integer characterizing the degree of movement of the system, and in general it is determined from a knowledge of the number of independent movements that the components would have if they were unconnected, together with the number and type of independent restrictions imposed on their movements by the interconnections. The section focuses on two-component systems, and we show how the constraints can be described most simply in terms of points in contact with surfaces. This is then related to the *connectivity* of the types of joint considered in Section 1.

In Section 4, *Planar kinematic systems,* we concentrate on a particular type of system — the planar kinematic system — in which the components move in parallel planes. After considering certain geometrical complications, we derive several *mobility criteria.* These are algebraic expressions determining the mobility of a planar system in terms of the numbers of components and joints. We end the section by considering the construction of a selection of planar systems with mobility 0 or mobility 1.

Section 5, *Flight simulators and robots,* is a television section. The television programme uses the examples of an aircraft flight simulator and a robot manipulator to discuss the design of kinematic systems for achieving the general movements of objects in three-dimensional space.

1 Kinematic structure

The general subject of kinematics deals with *movement* — a continuous change of position in space within a certain interval of time. In this unit we study mechanical systems which depend fundamentally on movement for their primary function. Such systems are referred to as *kinematic systems*.

The word *kinematics* is derived from the Greek κίνησις meaning a motion or movement; several other English words originate from the same root (for example, *cinema, kinetic, kinaesthetic* and *telekinesis*).

Definition

A **kinematic system** is a mechanical system designed to produce, transmit, control, constrain, or resist movement.

Here, rather than investigating the details of specific kinds of movement, we address the problem of the design of kinematic systems to achieve various *degrees* of movement. In particular, we investigate how to *arrange* and *interconnect* the component parts of a system so that their motions are suitably controlled and constrained. This is what is meant by **kinematic structure**.

This section deals with the fundamentals of kinematic structure and introduces the types of kinematic system that it is possible to design in two- and three-dimensional spaces.

1.1 Links and joints

Two simple examples of mechanical systems whose primary function depends crucially on kinematic design are scissors and door handles. Each consists essentially of a small number of component parts joined together in such a way that one component can rotate with respect to another. These two examples operate with a controlled *rotation*, but other simple kinematic systems involve different motions. For instance, bicycle pumps and retracting pens operate with a controlled straight-line movement (*translation*), whereas bottle caps and lids on jars operate with a controlled *screw motion*.

Other examples of kinematic systems are the focusing mechanisms on binoculars, telescopes and cameras and the pan-and-tilt mechanisms on tripods. Focusing mechanisms control a *relative sliding movement* between the tubes carrying the lenses, so that the distance between the lenses can be varied to bring the images into focus. Pan-and-tilt mechanisms allow *controlled rotational movements* enabling objects mounted on the tripod to be pointed in various directions.

Problem 1.1 ——————————————————————

List some examples of kinematic systems that are found in (or on) the following:

(a) a house;

(b) an aircraft;

(c) a yacht;

(d) a wheelchair;

(e) a computer.

More complicated mechanical systems whose primary function depends crucially on kinematic design are the internal-combustion engine and the robot manipulator.

In this section we give a wide range of examples. Do not worry if you are not familiar with them all.

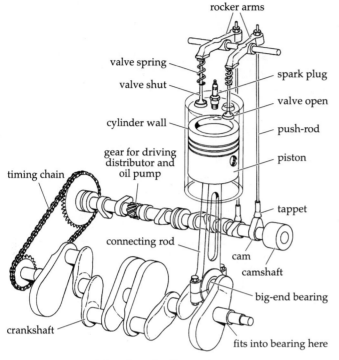

internal-combustion engine

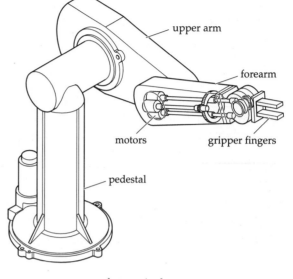

robot manipulator

Both of these systems are complex machines with many component parts, interconnected in a precisely defined geometric arrangement. There are *solid components,* most of which are essentially *rigid* (such as pistons, push-rods, crankshafts, gears, cams, arms and grippers), but some of which are *flexible* (such as chains, springs and hydraulic hosepipes). There are *fluid components* (such as lubrication oils, cooling fluids and hydraulic power transmissions) and there are *electrical* or *electronic components* (such as spark plugs, control circuits, generators and motors). How can we make sense of this plethora of detailed structure? Indeed, how can we even recognize the different components of a given system?

To simplify matters, we shall restrict our attention to rigid components and their (non-rigid) interconnections: we shall not consider flexible, fluid, electrical or electronic components. Furthermore, we shall not directly consider the detailed shapes and sizes of the parts of our system, but only the number and type of components and interconnections.

Definitions

The primitive rigid components of kinematic systems are called **links**.

The interconnections between links are called **joints**.

In common parlance, the word *link* is sometimes used to mean *joint,* so there is scope for confusion here. We suggest that you think of a link as if it were part of a chain, since this emphasizes that it is a rigid body.

In the internal-combustion engine system, the links are the pistons, connecting rods, crankshaft, cylinders, etc; the joints are the bearings, such as the big-end bearing, where the crankshaft and connecting rod are joined, or the plain bearing on the cylinder wall, where the piston and cylinder are in contact.

In the robot manipulator system, the links are the pedestal, upper arm, forearm, and gripper fingers; the joints are the connections between the upper arm and lower arm, etc. Because of the anthropomorphic nature of many robot manipulator systems, these joints are naturally referred to as shoulders, elbows, wrists, and so on.

Several examples of links and joints are to be found in familiar kinematic systems in the kitchen, such as: a cupboard door (link) and its hinge (joint); a cutlery drawer (link) and its runners (joint); a folding table-top (link) and its hinge (joint); and a pair of scissors — two blades (links) joined together at a pivot (joint).

Problem 1.2 ─────────────────────────────────────

List two examples of kinematic systems that are commonly found in:

(a) a home;

(b) an office;

(c) a factory;

(d) a playground;

(e) a department store.

In each case, choose a system where relative motion between two or more components is of *primary* importance in the design, and indicate some of the links and joints.

Types of link

So far, we have defined a *link* as a rigid body and a *joint* as a connection between links. As a first step in classifying links and joints, we need to be more precise than this in our definitions.

The simplest kinematic system consists of just *one* link which is free from all contact with other bodies (that is, there are no joints present). Such a link is called a **nullary link** since it has *no* joints associated with it. A familiar example of a nullary link is a pebble or stone tossed into the air and moving freely. The forces imposed by gravity and air resistance are not applied by direct contact with other solid bodies, so we do not consider them to be part of the system. In this sense we think of the stone as if it moved in a vacuum and beyond the reach of any gravitational or other forces. Nullary links, although apparently trivial, are important in establishing a correct combinatorial approach to kinematic design.

If we now introduce a second link, we obtain a kinematic system of two links. If these are not in contact at a joint, we just have a system of two nullary links. But, if the two links are joined by a single joint, each link has just *one* joint associated with it and each of the two links is referred to as a **unary link**. A familiar example of such a two-link system is provided by a piano — the piano body and the keyboard lid are two unary links joined at a single joint, the hinge.

Of course, in theory, the two links could be joined to each other by two (or more) joints, in which case neither link would be unary. However, in general, such a combination of joints would restrict (and probably prevent) any relative motion between the two links, unless the position and orientation of the joints satisfy special geometrical conditions. For example, consider a door and its frame. A door is almost always hung on more than one hinge, but the separate hinges can be considered to be part of one (long) hinge, since their geometrical arrangement gives them all the same axis of rotation and so they operate as a single unit.

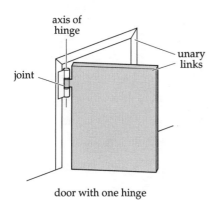
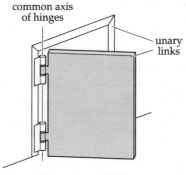

door with one hinge   door with two (coaxial) hinges

The introduction of a third link increases the possibilities further. We can now have three unconnected nullary links, or one nullary link together with two joined unary links. But, in addition, we can have one of the links connected to both of the other two, or each of the three links connected to both of the other two. In the former case, there is one **binary link** — that is, a link with *two* joints associated with it — connected to two unary links, forming a *serial arrangement*. In the latter case, there are three binary links interconnected in a *loop arrangement*.

A familiar example of a binary link in a serial arrangement occurs in a desk lamp. A binary link in a loop arrangement occurs in a kitchen balance (scales).

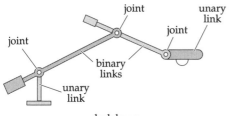
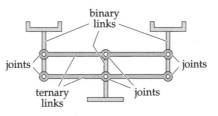

desk lamp
(binary links in serial arrangement)

kitchen balance
(binary links in loop arrangement)

We can continue in this way to define **ternary**, **quaternary** and, in general, **r-ary links**, which have respectively *three*, *four* and *r* joints associated with them. For example, the camera platform on an adjustable tripod is a ternary link, since it has three joints at which it is attached to the three legs, and the main body of a gate-leg table is a quaternary link, since it has four joints at which it is attached to the two drop leaves and the two gate legs.

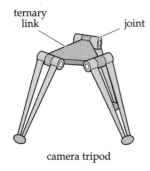
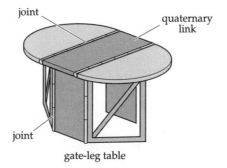

camera tripod   gate-leg table

Problem 1.3

List some common examples of each of the following types of link:

(a) unary; (b) binary; (c) ternary.

Table 1.1 illustrates some of the geometric forms that the various types of link can take. We depict the links schematically, using circles to denote potential sites for joints.

Table 1.1   Schematic forms of various types of link.

type of link	schematic forms	
	fundamental	some variations

In Table 1.1, the fundamental schematic form of each type of link represents an irregularly shaped body, with potential sites for joints located around the perimeter. The variations reflect the fact that the joints can be of different types, have different sizes and shapes, and be distributed in different ways, such as along a straight strip, around the edges of a polygon or, for links with four or more joints, at the corners of a polyhedron.

Some of the schematic forms can be misleading. For example, a ternary link drawn in strip or angle form, rather than in triangle form, can be mistaken for two binary links joined together. We could avoid this problem by always drawing a ternary link in trianglular form but this is often not convenient. Similar problems with quaternary links can be avoided by always drawing them in quadrilateral or tetrahedral form.

In all cases, the detailed shape of the link is of secondary importance, since we are mainly interested in the *number, type* and *arrangement* of joints on each link.

Types of joint

We now look in more detail at the concept of a *joint*. So far, we have defined a joint as a connection between links. We have implicitly assumed that the joint allows relative motion to take place between the links; we exclude such things as welded, glued or rigidly bolted connections.

The simplest type of joint is the connection of *two* links. Such a joint is called a **binary joint**. A familiar example is a door hinge.

If *three* different links are connected together by a single joint, the joint is called a **ternary joint**. With *four* links, we get a **quaternary joint**. In general, an **r-ary joint** is a single joint which connects *r* links.

Table 1.2 illustrates some of the geometric forms that the various types of joint can take. We depict each joint schematically, using a circle for the

There is no simpler joint than a binary joint, since the concept of a joint requires at least two links. It is meaningless to consider the notion of a 'nullary joint' (no links 'joined together') or a 'unary joint' (one link 'joined together'), since in neither case can any relative motion take place.

site of the joint and shaded areas radiating outwards to denote potential links meeting at the joint.

Table 1.2   Schematic forms of various types of joint.

type of joint	schematic forms	
	fundamental	some variations
binary		
ternary		
quaternary		
r-ary		

In Table 1.2, the fundamental schematic form of each type of joint represents a site where the appropriate number of links are connected together. In general, each of the links meeting at a joint can be unary, binary, ternary, ..., or r-ary, and we indicate this in the fundamental schematic forms by not specifying the shape of the shaded areas. The variations of schematic form allow for links to be drawn in strip, polygonal or polyhedral form.

You may be wondering why we have not presented a practical example of a ternary joint, or of an r-ary joint for any $r > 3$. Although we can sketch what appears to be an r-ary joint, it is difficult in practice to connect three or more links together at a single joint. On closer examination, such a joint is seen to consist of a collection of binary joints between *pairs* of links. For example, consider a threaded bolt on which two nuts (or a nut and a washer) are screwed. Ostensibly, the three links — the bolt and the two nuts — are joined at a single joint — the screw thread; but in fact the system consists of two separate binary joints — one between each nut and the bolt. Just as a ternary joint can be regarded as a combination of two binary joints, so a quaternary joint can be regarded as a combination of three binary joints. In general, an r-ary joint can be regarded as a combination of $r - 1$ binary joints.

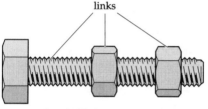

links

threaded bolt and two nuts

When r-ary joints ($r \geq 3$) are present in a system, they may conceal the true multiplicity of the links meeting at these joints. For example, in the following part of a system comprising three links meeting at a ternary joint, it appears that each link is in contact with each of the other two. This might lead us to interpret the single ternary joint as being equivalent to *three* binary joints at which the links meet. This interpretation is *incorrect*.

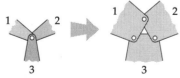

incorrect interpretation of ternary
joint as three binary joints

The main reason for rejecting this interpretation is that it means that the joint is inoperative — it has lost its freedom of movement — because we have created a triangle of links around a 'hole' and none of these links can move relative to the others. A correct interpretation of the ternary joint, which preserves the freedom of movement, is the following:

correct interpretation of ternary
joint as two binary joints

The single ternary joint is thus replaced by *two* binary joints in this process.

Problem 1.4

In how many correct ways can a ternary joint be expanded into binary joints?

This method of dealing with ternary joints in terms of binary joints can be generalized to accommodate other r-ary joints, and is useful for deriving appropriate mobility criteria in systems containing these joints. In general, an r-ary joint (at which r links meet) can be represented by an equivalent system of $r - 1$ binary joints, and this can be done in r^{r-2} different ways. Physically, we are replacing each r-ary joint with the minimum possible system of binary joints required to hold the links together and preserve the relative freedoms of movement.

This follows from Cayley's theorem for labelled trees (Theorem 2.1 of *Graphs 2*).

Sometimes it is useful to analyse a given kinematic system in terms of r-ary links and r-ary joints. However, it is usually more convenient (and indeed more realistic) to carry out the analysis in terms of r-ary links and *binary* joints. Therefore, for the rest of this unit,

we restrict our attention primarily to binary joints,

although we do make use of r-ary joints when these are more convenient. Binary joints are of such primary importance that much of kinematics can be based on their analysis. To emphasize the fact that binary joints involve just *two* links, we often refer to them as **kinematic pairs**.

Physically there are many possible types of kinematic pair, of different shapes, sizes and designs. Many are simple *pin* joints, allowing only a relative rotation between the two links which meet there. Others are *ball-and-socket* joints, allowing simultaneous relative rotations about more than one axis. Still others are *sliding* joints, which allow a relative translation between the two links. And there is the common *nut-and-bolt* joint, which allows a relative screwing motion between the two links.

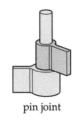

pin joint

ball-and-socket joint

sliding joint

nut-and-bolt joint

Problem 1.5

List some common everyday examples of kinematic pairs.

In Section 1.3 we examine the various types of kinematic pair in more detail, but first we show how to model kinematic structure using graphs.

1.2 Graph representations

Our eventual aim is to investigate techniques for enumerating all possible kinematic systems of a given class. To illustrate what we mean by this, consider a connected system containing just four links, and only binary joints. If we ignore the geometric aspects and consider just the possible interconnections, we can form just six different systems. These contain various types of link (ranging from unary to ternary), and various numbers of joints (ranging from three to six). We illustrate the systems in Table 1.3.

Table 1.3   The six connected kinematic systems with four links and only binary joints.

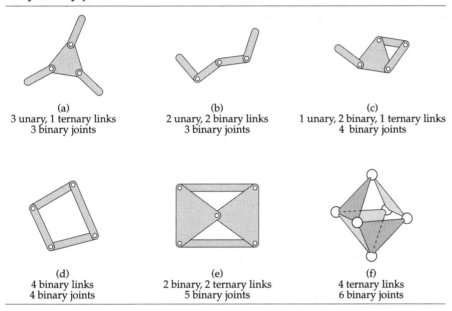

(a)	(b)	(c)
3 unary, 1 ternary links	2 unary, 2 binary links	1 unary, 2 binary, 1 ternary links
3 binary joints	3 binary joints	4 binary joints

(d)	(e)	(f)
4 binary links	2 binary, 2 ternary links	4 ternary links
4 binary joints	5 binary joints	6 binary joints

In order to control the internal motions of these systems, we need to control some of the joints (by driving them with electric motors, for instance). Systems (e) and (f) have no internal movement because they are completely constrained by the way that their links are connected, so none of their joints needs to be controlled: we say that they have *no mobility* (or zero mobility). However, each of the systems (a), (b), (c) and (d) has some internal movement: we say that they have *mobility* (or positive mobility). Later we examine the concept of mobility in more detail and define it to be an integer, but here we can make some progress without a rigorous definition.

An *internal* motion is one that changes the positions and orientations of the links of the system relative to one another. In contrast, an *external* motion (such as a translation or rotation) moves the entire system, relative to some frame of reference outside the system.

Problem 1.6

For each system in Table 1.3, state how many joints need to be controlled in order to control any internal motion.

The kinematic structure of each system in Table 1.3 can be formally represented in terms of graphs. The justification for this is two-fold. First, at the conceptual stage of design, a graph representation is useful in the enumeration of possible kinematic systems of a particular type. Second, it is important to have a precise representation of the kinematic structure of

This is helpful in determining possible patent infringements, for example.

a given system, so that any two systems can be compared for essential similarities or differences.

> ## Definitions
>
> The **direct graph** of a kinematic system is a graph in which the vertices represent the joints of the system and the edges represent the links of the system. In a direct graph, two vertices are joined by an edge if the corresponding joints belong to the same link.
>
> The **interchange graph** of a kinematic system is a graph in which the vertices represent the links of the system and the edges represent the joints of the system. In an interchange graph, two vertices are joined by an edge if the corresponding links are connected at a common joint.

direct graph
vertex ↔ joint
edge ↔ link

interchange graph
vertex ↔ link
edge ↔ joint

The *direct* graph of a kinematic system is so called because the graph looks like the system it represents. It is useful for representing systems containing multiple joints but only binary links, since the links are represented by edges. We present some examples in Table 1.4.

Table 1.4   Direct graph representations of some systems with binary links and multiple joints.

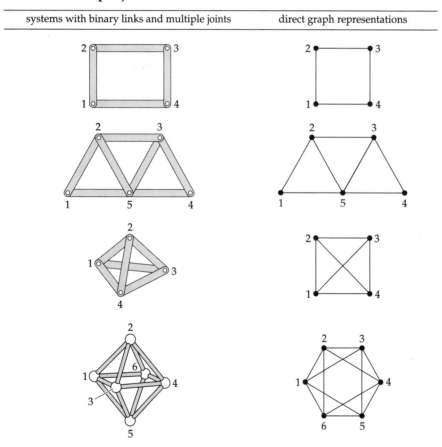

systems with binary links and multiple joints	direct graph representations

Problem 1.7

Explain why the direct graph representation is not useful for systems containing links that are not binary links. Suggest a possible extension of that representation that might be useful for such systems.

The *interchange* graph of a kinematic system is so called because the roles of the links and joints appear to be interchanged in the graph. It is useful for representing systems containing links other than just binary links but only binary joints, since the joints are represented by edges. We present some examples in Table 1.5.

Table 1.5   Interchange graph representations of some systems with binary joints and multiple links.

systems with binary joints and multiple links	interchange graph representations
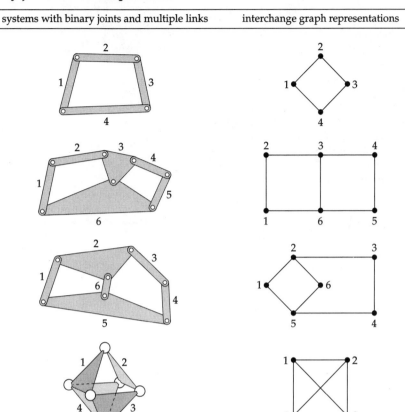	

Problem 1.8

What does the degree of a vertex represent in:

(a)   a direct graph?

(b)   an interchange graph?

Problem 1.9

Consider an interchange graph which is a connected graph, and which remains connected after the removal of any vertex. Show that, if it has n vertices and j edges, then the degree r of any vertex must satisfy the inequality

$$r \leq j - n + 2$$

We use this result in Section 4.5.

Hint   What happens to the graph if a vertex is removed?

Note that, for a system containing only binary links and only binary joints, the direct graph and the interchange graph are identical, although *they represent different things*.

system of four binary links
and four binary joints

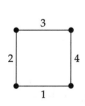

direct graph

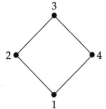

interchange graph

15

Also note that a given graph usually *represents two different systems* when considered first as a direct graph and then as an interchange graph.

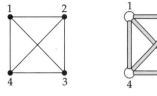
tetrahedron graph *G*

tetrahedral truss:
direct graph *G*

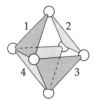
octahedral truss:
interchange graph *G*

In this unit we give more emphasis to interchange graphs than to direct graphs, because the restriction to only binary joints is less of a handicap than the restriction to only binary links. Indeed, as we have already noted that we can consider a multiple joint to be a combination of binary joints, there is no real restriction on the type of system that can be represented by an interchange graph.

We illustrate the usefulness of interchange graphs in the enumeration of possible kinematic systems by returning to the four-link systems of Table 1.3. We can enumerate the possible connected kinematic systems containing four links and only binary joints, by listing all simple *connected* graphs with four vertices. Table 1.6 shows these graphs, together with the corresponding systems.

Table 1.6   Interchange graphs of the six connected kinematic systems with four links and only binary joints.

interchange graph	corresponding system	interchange graph	corresponding system

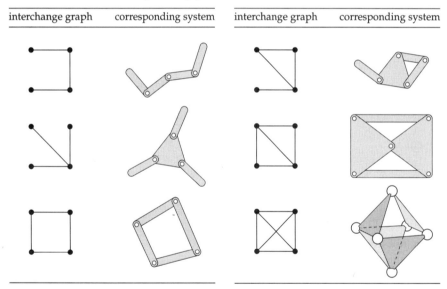

Problem 1.10

Sketch the six connected kinematic systems obtained by treating each of the six graphs in Table 1.6 as a direct graph representation.

1.3   Reuleaux pairs

It is possible to classify kinematic pairs (that is, binary joints) in a coherent scheme based on general aspects of the geometry of the connection, and on the relative motion that this permits.

For example, when we examine pin joints, ball-and-socket joints, sliding joints and nut-and-bolt joints closely, they reveal a **surface contact** between the two links involved. Thus the pin joint has cylinders and planes in contact, the ball-and-socket joint has spheres, the sliding joint has planes, and the nut-and-bolt joint has cylinders and helicoids (resembling a spiral ramp). These surfaces remain in contact throughout the motion, and thereby constrain the links to move in the required way.

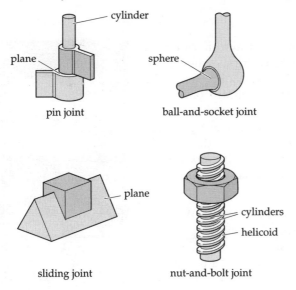

In addition to these surface-contact kinematic pairs, there are types of kinematic pair in which there is only **line contact**. An example of this type of kinematic pair is provided by a rocking chair in contact with the ground (two coincident contact lines) or by a pair of meshing gears (one contact line).

Although we refer to *line* contact here, and to *point* contact below, these are idealizations, since any real physical system would deform under a load to produce a small area of contact (the contact patch).

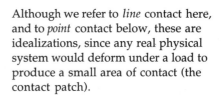

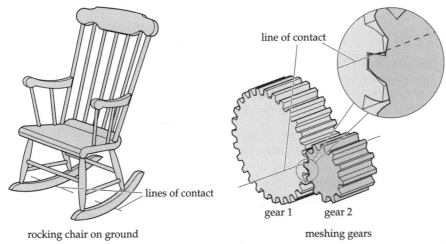

rocking chair on ground

meshing gears

Similarly, there are kinematic pairs in which the two links are in contact only at a finite set of discrete points. An example of this **point contact** is provided by a ball-point pen writing on a page (one contact point) or a tripod standing on the ground (three contact points).

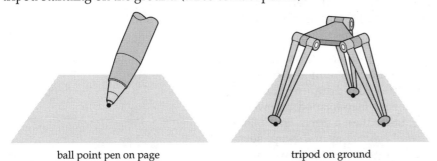

ball point pen on page

tripod on ground

We therefore see that two links can be connected in three fundamentally different ways to form a kinematic pair: they may have

- a finite number of *points* in common;
- a finite number of *lines* in common;
- a finite number of *surfaces* in common.

In the nineteenth century, the kinematician Franz Reuleaux investigated binary joints in his attempts to derive a classification scheme for machines. He distinguished two basic classes of binary joint, which he referred to as **higher kinematic pairs** and **lower kinematic pairs**. His classification is much the same as ours, except that he grouped the point-contact and line-contact joints together as *higher* pairs. His *lower* pairs are essentially the same as our surface-contact joints. Reuleaux defined six lower kinematic pairs having surface contact. Each of the six allows different relative movements between the two links forming the kinematic pair.

We shall not discuss higher kinematic pairs further in this unit. We shall instead concentrate on the six Reuleaux lower kinematic pairs — commonly referred to as the **Reuleaux pair**s — since they are the most important surface-contact binary joints, and all other types can be derived from them. They are illustrated in Table 1.7, together with their associated contact surfaces. Before we discuss each of them in detail, we introduce some terminology.

FRANZ REULEAUX (1829–1905)

Franz Reuleaux studied in Karlsruhe, Berlin and Bonn. He was appointed professor of machine design at the polytechnic school in Zurich at the age of twenty-seven, and was later made director of the Royal Industrial Academy in Berlin in 1868. He helped to found the Royal Institute of Technology in Berlin, and served as professor there until 1896.

Definitions

The combined position and orientation of a link in space, relative to a reference coordinate system, is called its **pose**. If the pose changes with time, then the link is in **motion**.

We now examine each of the Reuleaux pairs in turn. They are distinguished from one another by the type of movement of one of the links in the pair with respect to a coordinate system fixed in the other link. The precise details of their mechanical construction are of secondary importance in the classification of the six types.

For the **revolute pair**, one of the links is free to *rotate* about an axis, fixed with respect to the other link, but cannot move in any other way. Its pose throughout its motion is specified by a *single* quantity — the angle through which it has turned. The contact surface for a revolute pair can be any surface of revolution about the axis of motion, except a cylinder or sphere. We use a *cone* as our surface of revolution, since this is the simplest such surface. It is generated by revolving a straight line (which intersects the axis in a single point) about that axis. A common example of the revolute pair is provided by a hinge, although usually the contact surface is then a combination of a cylinder (to allow rotation about the hinge axis) with two planes (to prevent sliding along the hinge axis). The cylinder in a hinge usually has a small diameter relative to its length and so resembles a pin — hence the common alternative name *pin joint* for a revolute pair.

For the **prismatic pair**, one of the links is free to *translate* along an axis, fixed with respect to the other link, but it cannot move in any other way. Its pose is also specified by a *single* quantity — the distance moved. The contact surface for a prismatic pair can be a surface formed by sweeping along the axis of motion any plane curve, other than a circle, in a plane at right angles to that axis. We use any *elliptical cylinder* as our swept surface, since this is the simplest such surface. It is generated by sweeping an ellipse along the axis of motion. An elliptical cylinder allows translation without rotation, whereas a circular cylinder would allow both. A common example of the prismatic pair is provided by a *sliding joint* such as a curtain rail and glider, although usually the contact surface is then a triangular or rectangular prism (hence the name *prismatic*).

The **screw pair** permits one of the links to move in a combination of a *rotation* about and a *translation* along an axis fixed with respect to the other link. However, the two motions are coupled together, so that only *one* quantity — such as the angle turned through — is required to specify the pose of one link relative to the other. The contact surface for a screw pair can be any surface formed by sweeping a plane curve along a helical path whose axis is the axis of motion. We use a *helicoid* as our swept surface, since this is the simplest such surface. It is generated by a coupled translation and rotation of a straight line (which intersects the axis at right angles) along that axis. A common example of the screw pair is provided by a nut and bolt, although usually the contact surface is then a screw thread (hence the name *screw*) which is a more complicated surface than a simple helicoid.

For the three remaining pairs, the situation is more complicated, since each requires more than one quantity to specify the pose of one link with respect to the other.

Table 1.7 The six Reuleaux pairs.

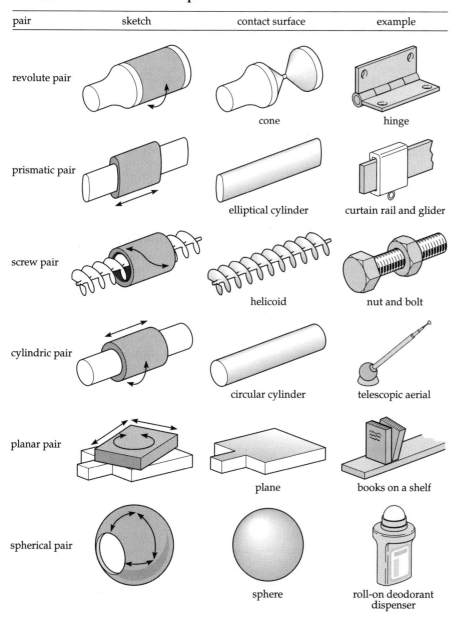

pair	sketch	contact surface	example
revolute pair		cone	hinge
prismatic pair		elliptical cylinder	curtain rail and glider
screw pair		helicoid	nut and bolt
cylindric pair		circular cylinder	telescopic aerial
planar pair		plane	books on a shelf
spherical pair		sphere	roll-on deodorant dispenser

The **cylindric pair** permits *independent* rotation about and translation along an axis, so that *two* quantities — an angle and a distance — are required to specify the pose of one link relative to the other. The contact surface for a cylindric pair is a *circular cylinder* (hence the name *cylindric*) whose axis is the axis of motion. A common example of the cylindric pair is provided by two adjacent sections of a telescopic aerial.

The **planar pair** permits relative translations in two independent directions on a plane surface. In addition, relative rotation about a line perpendicular to this surface is possible, so that the planar pair requires *three* quantities — two distances and an angle — to specify the pose of one link relative to the other. The contact surface for a planar pair is a *plane* (hence the name *planar*). A common example of the planar pair is provided by two adjacent books on a bookshelf, as can be seen by observing the rotation and translations as one book is retrieved from between two others on the shelf.

The **spherical pair** permits relative rotations about three independent axes passing through a single point. So the spherical pair requires *three* quantities — three angles — to specify the pose of one link relative to the other. The contact surface for a spherical pair is a *sphere* (hence the name *spherical*). A common example of the spherical pair is provided by a roll-on deodorant dispenser (a type of ball-and-socket joint).

Problem 1.11 ───────────────────────────────

By referring to Table 1.7, give one further example of each of the six Reuleaux pairs.

The most fundamental of the six Reuleaux pairs is the screw pair, since it is possible to derive the remaining five pairs from just this one. To see how this is done, we need to compare the relative motions between the two links forming each Reuleaux pair, since this is essentially what distinguishes the six types.

Consider the screw pair, and notice the motion of a point on the 'nut' as this body moves along the screw threads. The point traces out a definite curve in space, and the curve is a *helix* having the same *pitch* as the screw threads.

The *pitch* of a helix is the distance between corresponding points on successive coils.

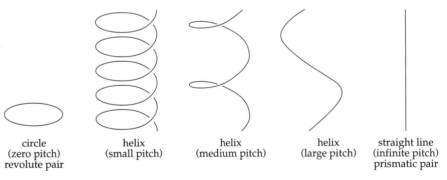

circle	helix	helix	helix	straight line
(zero pitch)	(small pitch)	(medium pitch)	(large pitch)	(infinite pitch)
revolute pair				prismatic pair

helices corresponding to screw pairs

If the pitch of the helix is small, then the nut does not advance very far along the axis as it rotates through a given angle. If the pitch is large, then the nut advances a large distance along the axis as it rotates through the same given angle. There are two extreme cases. In one, the helix has *zero pitch*, so that it collapses into a circle, and the nut can only rotate about the axis without advancing at all. In the other, the helix has *infinite pitch*, so that it is stretched out into a straight line, and the nut cannot rotate at all but can only advance along its axis. So, a screw pair with zero pitch allows a relative motion which is identical to that of the revolute pair (a rotation), whereas a screw pair with infinite pitch allows the same relative motion as a prismatic pair (a translation). Thus, the revolute and prismatic pairs are special types of screw pair.

Furthermore, the remaining three pairs (the cylindric, the planar and the spherical pairs) can all be simulated by combinations of two or three screw pairs (possibly with zero or infinite pitch).

Table 1.8 Combinations of Reuleaux pairs.

Reuleaux pair	sketch	equivalent combinations
cylindric pair		• one revolute pair and one prismatic pair • two screw pairs
planar pair		• two prismatic pairs and one revolute pair • three revolute pairs
spherical pair		• three revolute pairs

For instance, a combination of a revolute pair and a prismatic pair sharing the same axis can be substituted for the cylindric pair. The substitution process introduces a third, intermediate, link into the system, but the relative motion of the two outer links in the combination remains that of a cylindric pair.

Table 1.8 lists some combinations of screw pairs for those Reuleaux pairs requiring more than one quantity to specify their pose.

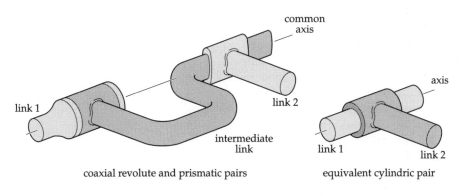

coaxial revolute and prismatic pairs   equivalent cylindric pair

Problem 1.12 ───────────────────────

By referring to Tables 1.7 and 1.8, and using the style of the Figure above, sketch the geometrical arrangement of axes of revolute, prismatic and screw combinations for each of the following:

(a) a cylindric pair simulated by two screw pairs;

(b) a planar pair simulated by two prismatic pairs and one revolute pair;

(c) a spherical pair simulated by three revolute pairs.

Hint Think carefully about the relationships between the various axes.

The screw pair thus plays a central role as the most fundamental type of binary joint with surface contact. We could therefore base the whole subject of kinematic structure on this one type of joint. However, it is usually more convenient to treat the six Reuleaux pairs as different types of binary joint.

1.4 Kinematic geometry

So far we have avoided a discussion of the detailed geometry of kinematic systems and have focused on the interconnections of the links. We continue with this approach for the remainder of this unit, but first we briefly examine some important geometrical aspects that have a strong influence on kinematic structure. Specifically, the juxtaposition and relative orientation of the Reuleaux pairs in a kinematic system are crucial in determining the type of motion possible.

We begin by considering systems containing several kinematic links interconnected at various kinematic joints. We shall not discuss these systems at great length here, because we deal specifically with them later, in Sections 3 and 4. For the moment we briefly discuss some aspects of their kinematic geometry, in terms of which the systems are classified.

Let us examine a four-link system consisting of four binary links connected in a loop by four revolute pairs. An example of this kind of system is provided by the door, hinge, frame and closing mechanism of a self-closing door.

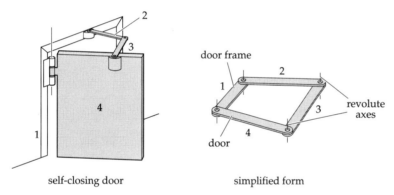

self-closing door   simplified form

The system has two more links and three more revolute pairs than a simple door, together with a spring to apply the closing force and a damper to control the rate of motion. If we ignore the spring and damper (since these do not affect the kinematic structure), then the system is essentially just four binary links interconnected by four binary joints to form a simple closed quadrilateral loop. Notice, however, that the detailed geometry of the system is rather special, in that the four revolute pairs all have parallel axes. In general, this need not happen; the axes of the joints in *any* system (closed quadrilateral loop or otherwise) may, for example, all intersect in one point, or they may all be skew to one another (that is, no two intersect or are parallel).

These possibilities are illustrated in Table 1.9.

The three special cases we have singled out are important since they occur widely in many types of machine, and they are given special names. When all of the axes are parallel, we refer to the system as a **planar system**, because the links all move in parallel planes. When all of the axes intersect in a single point, the system is called a **spherical system**, because the links all move on the surfaces of concentric spheres. Finally, if the axes are all mutually skew, we call the system a **spatial system**, because all the links have general motions in space.

Other possibilities exist: for example, two or three of the axes may be parallel or pairs of axes may intersect in a point, and so on.

These observations form the basis of a general classification scheme for kinematic systems. Table 1.9 illustrates the idea with simple examples.

Table 1.9   Planar, spatial and spherical kinematic systems.

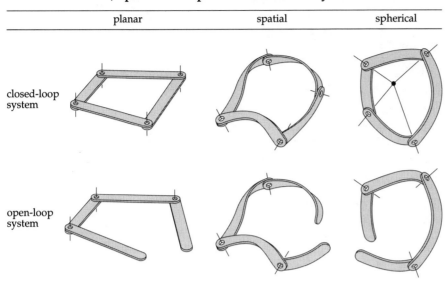

	planar	spatial	spherical
closed-loop system			
open-loop system			

Whether or not a kinematic system is planar, spherical or spatial, it can have many links and joints of various multiplicities (unary, binary, ternary, quaternary, and so on). The links can be connected in a closed loop, so that the interchange graph is a cycle, or the system may have no loops, so that the interchange graph is a tree. In general, a system has a combination of loops and branches, and the interchange graph has both cycles and branches; we then refer to **multi-loop/multi-branch kinematic systems**. An example is the set-square mechanism found on drawing boards.

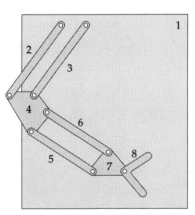

set-square mechanism

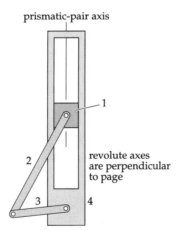

slider-crank mechanism

prismatic-pair axis

revolute axes are perpendicular to page

The numbers refer to the links.

In planar kinematic systems, the joints may be revolute pairs (provided all the revolute axes are parallel), prismatic pairs (provided that all the prismatic axes are at right angles to the revolute axes, and hence are parallel to the plane of motion) or planar pairs (provided that the plane of contact is at right angles to the revolute axes and parallel to the prismatic axes). The links then move in parallel planes. An example of a planar kinematic system with both revolute and prismatic pairs is the slider-crank mechanism, commonly found in engines and pumps.

In spherical kinematic systems, we permit only revolute pairs, since on a sphere the equivalent of a prismatic pair (allowing sliding over the surface) is really just another revolute pair.

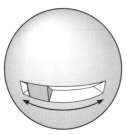 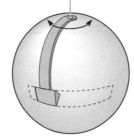

'prismatic' pair on a sphere   equivalent revolute pair on a sphere

The most familiar example of a spherical system is the *Hooke's joint* — a type of universal joint used as a coupling between two non-parallel rotating shafts. There are at least two Hooke's joints on the transmission shafts of most cars, and they can often be seen on trucks and lorries, where they are usually exposed.

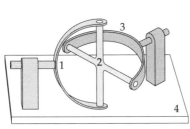
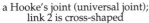
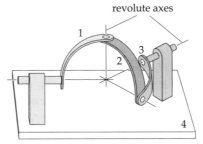

a Hooke's joint (universal joint); link 2 is cross-shaped

equivalent form showing the four intersecting revolute axes

The design of Hooke's joint most often found does not appear at first sight to be a spherical system, because it contains a cross-shaped link not lying on the surface of the sphere. This link provides mechanical strength, although from the point of view of kinematic structure it is equivalent to a simple binary link on the surface of the sphere. The system is a spherical kinematic system because the four revolute axes all intersect in the same point throughout the motion.

Spatial systems are the most general type of kinematic system and in such systems all six Reuleaux pairs can be present, although revolute, prismatic and spherical pairs are perhaps the most commonly occurring. For example, a desk lamp (of the Anglepoise type) is a spatial kinematic system with one loop, two branches and all revolute pairs.

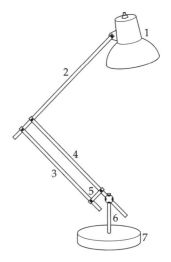

Spatial kinematic systems are in widespread use in the form of robot manipulators, which are often open-loop systems. These frequently contain prismatic pairs (in the form of hydraulic rams or booms), revolute pairs (in the form of actuating electric or hydraulic motors), and screw pairs (in the form of lead screws). In addition to the robot manipulator, another spatial kinematic system that has recently become familiar is the flight simulator.

Spatial kinematic systems containing just spherical pairs are commonly used to model structural frameworks. Here the system must resist motion in the most efficient way (for example, by being as light and as rigid as possible), and the function of the spherical pairs is to accommodate the small strain movements set up at the joints by the stresses caused by the weight of the structure. A welded joint will not absorb the strain in this way, and may be overstressed (that is, liable to fracture or failure) as a result.

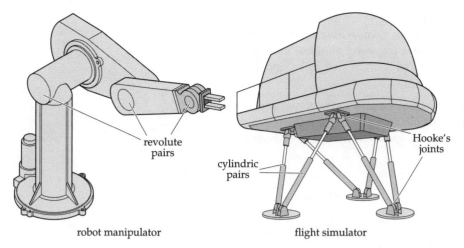

revolute
pairs

Hooke's
joints

cylindric
pairs

robot manipulator

flight simulator

We examine these systems in
Section 5.

Problem 1.13

Give an example of a spatial kinematic system found on a building site.

The classification of kinematic systems into planar, spherical or spatial types is important in determining the amount or degree of flexibility of motion that they possess. We use the term *mobility* to refer to the possible motions of a system, after we have taken account of the restrictions imposed by the interconnections of the links.

Definitions

The **mobility** of a kinematic system is the number of independent quantities or coordinates that must be specified in order to describe the pose of every link of the system with respect to a coordinate frame fixed on a chosen reference link, called the **fixed link**.

Consider the simple example of the door, hinge and frame. It is natural to choose the frame as the fixed link, and only one quantity is needed to describe the pose of the other link — the door — relative to the fixed link, namely the angle the door makes with the frame. So the mobility of this system is 1.

Let us return now to our example of the self-closing door, and determine its mobility as a kinematic system. Again, it is natural to choose the frame as the fixed link, and only one quantity is needed to describe the poses of all the links in the system relative to the fixed link, namely, the angle the door makes with the frame, so the mobility is 1. You may be surprised that the mobility is so small, since there are four binary links and four binary joints (revolute pairs) present, and yet the system has the same mobility as a system of two unary links and one binary joint (that is, the door, hinge and frame). However, the result is easily verified, since the motion of the door is of the same type as it was before, a simple rotation specified by an angle, and the positions of the extra links are entirely determined by the door's position. The additional motions introduced by the extra links and joints have been controlled by the arrangement of the components into a closed-loop system.

The above examples raise the following questions. How can we determine the mobility of a kinematic system? Can we do this in terms of the number and type of links and the number and type of joints, or do we require more information, such as the number of loops or the geometrical arrangement of the joint axes?

The subject of mobility occupies us for much of the remainder of this unit. We classify kinematic systems in terms of their mobility, and then

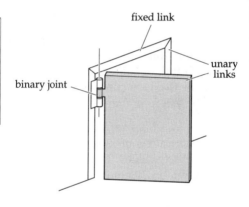

fixed link

binary joint

unary
links

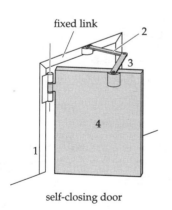

fixed link

2

3

4

1

self-closing door

attempt to enumerate all possible types with a given mobility and a given number of links. Our classification depends on a knowledge of the motions of the individual links and on the restrictions imposed on these motions by the various types of joint, as well as on the number of links and joints present in the system.

Mobility also depends on the geometry, since certain geometrical arrangements can introduce, restrict or remove motions. An example of this is provided by the four-link closed-loop kinematic systems shown in Table 1.9.

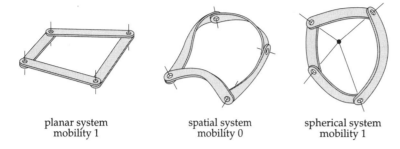

planar system	spatial system	spherical system
mobility 1	mobility 0	mobility 1

If the axes are all parallel, we have the planar system with mobility 1 — the four-link single-loop planar kinematic system exemplified by the self-closing door.

Similarly, if the axes all intersect in a single point, we have the spherical system, and this also has mobility 1, since specifying just one joint angle determines the pose of each link.

However, in the spatial system, all the revolute axes are skew, and the system does *not* move — it is locked in position. The individual motions of each link of the system are not compatible — they cannot move in unison since they are constrained to move in different directions. The mobility of the system is therefore 0.

We shall generally avoid such geometrical complications in our classification of kinematic systems by confining our attention to particular types of system. Thus, for example, in Section 4, where we examine *planar* kinematic systems, we take account of the special geometry of parallel revolute pair axes by considering the links to move only in two dimensions (parallel to a plane) rather than in three dimensions.

After studying this section, you should be able to:

- recognize a *kinematic system* and give some common examples;

- explain the terms *kinematic structure, link* and *joint*;

- distinguish the various types of link and joint in terms of their multiplicity;

- describe the two types of graph representation of a kinematic system — the *direct graph* and the *interchange graph* — and construct such a graph for an appropriate given system;

- understand the difference between *higher kinematic pairs* and *lower kinematic pairs* in terms of the type of contact;

- describe the six *Reuleaux pairs*;

- distinguish between *planar, spatial* and *spherical kinematic systems*;

- explain the terms *pose, motion* and *mobility*.

2 Braced rectangular frameworks

In this section we consider a kinematic design problem encountered in structural engineering. Many buildings are supported by steel frameworks, which usually consist of rectangular arrangements of girder beams and welded or riveted joints. This is particularly the case for high-rise buildings such as the so-called skyscrapers, skyprickers and skycities. Now you may think that architects always design buildings so that no movement can occur; but, with such tall structures, small movements are unavoidable (and often desirable) under stresses such as those caused by wind forces. Moreover, in earthquake zones, such as Japan or the west coast of the USA, buildings must be designed to yield to such forces in a controlled and predictable manner so that their motions damp out any resonance and prevent catastrophic collapse. Indeed, the motion of tall buildings is sometimes controlled dynamically with heavy pendulums or hydraulic rams oscillating very large masses continuously through small distances inside the building.

This happens in the Hancock building in Boston, Massachussetts, which isn't in an earthquake zone but is subjected to strong turbulent winds.

In each of these cases, a simple *rectangular* framework by itself is inadequate as a stable support structure. A rectangular framework maintains its shape by having rigid joints at its corners, and if these fail (often under relatively small loads) the rectangle deforms by bending at its corners to form a parallelogram. Its *geometry* is inherently unstable.

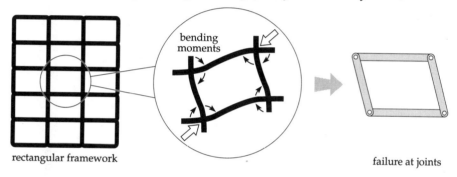

rectangular framework

bending moments

failure at joints

In contrast, a *triangular* framework strongly resists any changes in its shape, and fails (only under large loads) when its sides buckle or its joints separate. Its *geometry* is inherently stable. Indeed, its joints need not be rigid, and may be pin joints (revolute pairs) without much loss of strength. This is why so many structures, such as bridges and roof trusses, are triangulated.

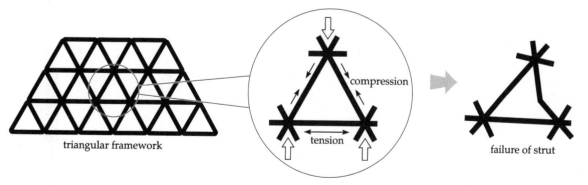

triangular framework

compression

tension

failure of strut

Despite the inherent lack of stability in rectangular frameworks, they are widely used, because they fit the natural bias towards the horizontal, the vertical and the perpendicular in buildings. Thus they provide strong vertical columns to support the weight of the building, and strong horizontal beams to support the concrete floors.

In this section we address the problem of making such rectangular frameworks stable by converting them into essentially triangular frameworks.

2.1 Braces

A rectangular framework consists of one or more rectangles, or **bays**. If we treat the corners of a bay as kinematic joints rather than welded joints, each bay can be modelled as a planar kinematic system consisting of four binary links connected in a closed-loop arrangement by four revolute pairs. We show below how such a bay moves when considered in isolation.

It is similar to the self-closing door considered in Section 1.

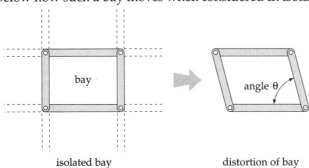

isolated bay   distortion of bay

In Section 1.4 we found that the mobility of this system is 1, because a single quantity, such as the angle θ between two adjacent binary links, specifies the pose of all four links. So, if we wish to prevent it from moving, we must somehow fix this angle. The most obvious way to do this is to weld one of the joints, in which case we essentially combine two adjacent binary links into one link. This reduces the system to one consisting of three binary links connected by three revolute pairs to form a triangle: the rectangular bay becomes a triangular bay.

rectangular bay with one welded joint   equivalent triangular bay

Although this method stabilizes the bay, the welded joint can be highly stressed and liable to failure. So, more robust alternatives are preferred in practice. We could fix the angle at a corner of the bay in one of the following ways: (a) adding an extra link to join any two adjacent links; (b) adding an extra link to join any two opposite links; (c) adding a link across either diagonal; (d) inserting a rectangular plate into the bay.

(a) link across adjacent links   (b) link across opposite links

(c) link across diagonal   (d) rectangular plate inserted

Problem 2.1

State the number and type of links and joints in each case after the bay is stabilized by:

(a) a link across two adjacent links;

(b) a link across two opposite links;

(c) a link across a diagonal.

Each of the five ways of stabilizing the bay we have considered reduces the mobility from 1 to 0, so they are all equivalent in this respect.

Definitions

A **brace** is a mechanical restriction on the motion of a bay in a rectangular framework which reduces the bay's mobility by 1, so that one fewer quantity is required to specify the pose of every link of the framework. A **bracing** of a framework is a particular allocation of braces to bays of the framework.

We have seen five different ways of implementing a brace. From now on we shall consider a brace to be a rectangular plate, regardless of how it is constructed in practice, since this leads to uncluttered diagrams. We indicate the insertion of a plate by shading the appropriate bay in the framework, as in the *Introduction* unit.

The insertion of a brace into a bay in a rectangular framework essentially combines all four binary links (the edges of the bay) into a single link (the boundary and interior of the bay). But what type of link? It depends where the bay is located in the framework. If the framework consists of just one bay, then a brace turns the bay into a nullary link. If the framework is just a string of bays, then bracing an *end* bay turns the bay into a *binary* link, whereas bracing an *internal* bay produces a *quaternary* link.

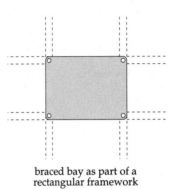

braced bay as part of a
rectangular framework

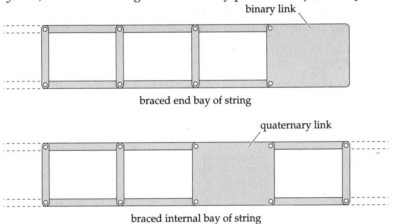

braced end bay of string

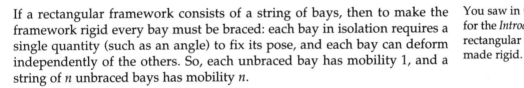

braced internal bay of string

If a rectangular framework consists of a string of bays, then to make the framework rigid every bay must be braced: each bay in isolation requires a single quantity (such as an angle) to fix its pose, and each bay can deform independently of the others. So, each unbraced bay has mobility 1, and a string of n unbraced bays has mobility n.

You saw in the computing activities for the *Introduction* unit how a simple rectangular framework of this type is made rigid.

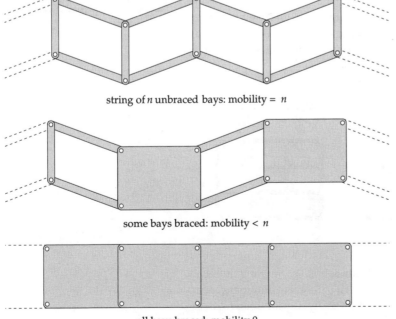

string of n unbraced bays: mobility = n

some bays braced: mobility < n

all bays braced: mobility 0

29

Notice that the bays in such a framework can have different sizes as long as they are compatible with one another — that is, each pair of adjacent bays shares a common edge. This holds for more complicated situations where the framework extends in two directions, and in such a case each bay must be compatible with its neighbours in both the horizontal and vertical directions. However, without any loss of generality, we shall simplify matters by adopting a *standard rectangular framework* in which all the bays have the same size and (undeformed) shape — that is, in which all the bays are congruent rectangles.

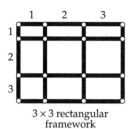
3 × 3 rectangular framework

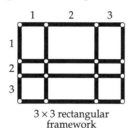
3 × 3 rectangular framework

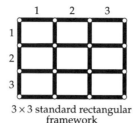
3 × 3 standard rectangular framework

equivalent 3 × 3 rectangular frameworks

In general, a large rectangular framework has high mobility, since many coordinates are needed to specify its pose. However, it is not necessary to brace every bay of such a framework in order to reduce its mobility to zero — that is, to make the framework *rigid*.

This contrasts with the situation of a framework comprising a string of bays, where every bay does need to be braced.

rigid

non-rigid

So, given a general rectangular framework, which bays need to be braced in order to make it rigid? Before we can begin to answer this question, we introduce some terminology.

Definitions

In a standard rectangular framework, a **row** is the set of all links forming the vertical sides of a horizontal string of bays, and a **column** is the set of all links forming the horizontal sides of a vertical string of bays.

If the rows are numbered $r_1, r_2, ..., r_i, ..., r_n$ sequentially from top to bottom and the columns are numbered $c_1, c_2, ..., c_j, ..., c_m$ sequentially from left to right, then **bay(i, j)** is the bay whose vertical links belong to row r_i and whose horizontal links belong to column c_j.

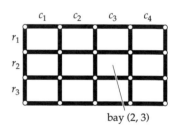
bay (2, 3)

Consider now the three frameworks shown below. Framework (a) has six bays braced, whereas each of the other two has just five bays braced.

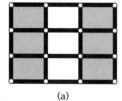
(a)

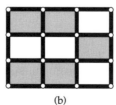
(b)

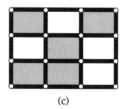
(c)

Only one of the three frameworks is rigid — framework (c) — even though it has fewer braces than the first. Although all three frameworks are braced, frameworks (a) and (b) deform kinematically under load, since

30

neither is braced in a way that reduces its mobility to zero; they deform as shown below.

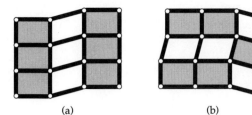

(a)   (b)

Framework (a) has mobility 1, and has a complete column unbraced. Is this the reason why it is not rigid? Must we have at least one brace in each row and in each column? Well, framework (b) has just such a bracing, but it is not rigid, so the criterion for rigidity cannot be as simple as this.

Let us take a closer look at what happens to the links when a framework deforms. Under a kinematic deformation, all of the links in any row remain parallel. Similarly, all of the links in any column remain parallel. Also, if a bay in a framework is braced, then all of the links in this bay's row are perpendicular to all of the links in this bay's column. However, in general, the links in other rows and columns do not satisfy either the parallelism or the perpendicularity relationships with the links of the braced bay.

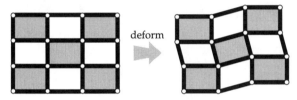

deform

Now, if a framework is to be rigid, we must at least ensure that *all* the links in *all* the rows must always remain parallel and that *all* the links in *all* the columns must also always remain parallel. We can use our observations about what happens to the links in a bay's row and column to deduce whether these properties hold for any given framework. For example, consider framework (c). From the observation that all of the links in a braced bay's row must remain perpendicular to all of the links in the bay's column, even after any possible deformation of the framework, we can deduce that:

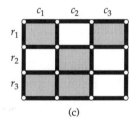

(c)

    row r_1 must remain perpendicular to columns c_1 and c_3
    row r_2 must remain perpendicular to column c_2
    row r_3 must remain perpendicular to columns c_1 and c_2

Now, since column c_1 must remain perpendicular to rows r_1 and r_3, we can deduce that rows r_1 and r_3 must remain parallel to each other. Similarly, we can deduce that rows r_2 and r_3 must remain parallel to each other, since they both must remain perpendicular to column c_2. Hence all three rows must remain parallel to each other. Consequently, since row r_1 must remain perpendicular to columns c_1 and c_3 and since row r_2 must remain perpendicular to column c_2, we can deduce that all three columns must remain parallel to each other. Furthermore, we can deduce that each row must remain perpendicular to each column. The only possible conclusion arising from the facts that all three rows must remain parallel to each other, all three columns must remain parallel to each other, and each row must remain perpendicular to each column, is that the framework is rigid.

This discussion leads us to be able to deduce the following general criterion for a braced rectangular framework to be rigid.

Theorem 2.1

A braced rectangular framework, with rows $r_1, r_2, \ldots, r_n$ and columns $c_1, c_2, \ldots, c_m$ is rigid if and only if its braces are located such that, for $i = 1, 2, \ldots, n$ and $j = 1, 2, \ldots, m$:

(a) r_i must remain parallel to r_j, for all r_i and r_j

(b) c_i must remain parallel to c_j, for all c_i and c_j

(c) r_i must remain perpendicular to c_j, for all r_i and c_j

under any attempted deformation of the framework.

Problem 2.2

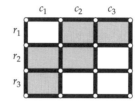

Verify that conditions (a), (b) and (c) of Theorem 2.1 are satisfied for the braced rectangular framework shown in the margin, and deduce that the framework is rigid.

The braced rectangular framework considered in Problem 2.2 satisfies the conditions of Theorem 2.1 and hence is rigid. But the task of checking the conditions becomes very tedious as the number of rows and columns in the framework increases significantly. Is there a better approach?

2.2 Braced frameworks and bipartite graphs

The conditions stated in Theorem 2.1 can be formulated in terms of a bipartite graph. This graph formulation provides a powerful technique for determining the rigidity of a given rectangular framework, and for deciding where to insert braces. In addition, it provides a systematic method for obtaining a rigid framework which is *minimally* braced, in the sense that it has the smallest possible number of braces, which in practice can lead to significant savings in materials when building framework-based structures.

To see how it works, consider the 3×3 framework in Problem 2.2. We model this with a bipartite graph whose two sets of vertices, labelled r_1, r_2, r_3 and c_1, c_2, c_3, represent the rows and the columns of the framework respectively. A vertex r_i from the row set is joined by an edge to a vertex c_j from the column set if and only if $\text{bay}(i, j)$ is braced.

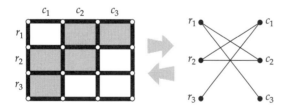

Problem 2.3

Draw the bipartite graph corresponding to each of the following braced frameworks:

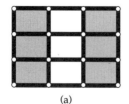

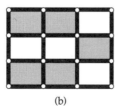

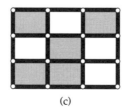

(a)   (b)   (c)

These are the frameworks (a), (b) and (c) that you saw earlier.

Which of the bipartite graphs are connected?

Framework (c) in Problem 2.3 is rigid and its bipartite graph is connected. This is also the case for the framework in Problem 2.2. However, frameworks (a) and (b) in Problem 2.3 are not rigid and their bipartite graphs are not connected. So, is a similar relationship true in general? Does the connectedness of the bipartite graph correspond to the rigidity of the framework?

Consider the framework from Problem 2.2 together with its bipartite graph.

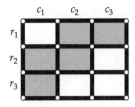

in the framework		*in the graph*
bay(1,2) is braced, so row r_1 must remain perpendicular to column c_2	$\Rightarrow$	r_1 is joined to c_2 by an edge
and		and
bay(2,2) is braced, so row r_2 must remain perpendicular to column c_2	$\Rightarrow$	r_2 is joined to c_2 by an edge
$\Downarrow$		$\Downarrow$
row r_1 must remain parallel to row r_2	$\Rightarrow$	r_1 is joined to r_2 by the path $r_1 c_2 r_2$

This correspondence between parallelism and perpendicularity in the framework and a path in the graph extends to all pairs of rows and/or columns which must remain parallel or perpendicular (because of braces in the framework). Furthermore, the correspondence enables us to deduce the maintenance of parallelism or perpendicularity in the framework from any path in the graph. For example:

in the framework		*in the graph*
		r_3 is joined to r_2 by the path $r_3 c_1 r_2$
		$\Downarrow$
row r_3 must remain perpendicular to column c_1	$\Leftarrow$	r_3 is joined to c_1 by an edge
and		and
row r_2 must remain perpendicular to column c_1	$\Leftarrow$	r_2 is joined to c_1 by an edge
$\Downarrow$		
row r_3 must remain parallel to row r_2		

Problem 2.4

The examples above enable us to list the following correspondences between rows and/or columns that must remain parallel and/or perpendicular in the framework of Problem 2.2 and paths in its bipartite graph:

r_1 parallel to r_2 $\leftrightarrow$ $r_1c_2r_2$
r_3 parallel to r_2 $\leftrightarrow$ $r_3c_1r_2$

r_1 perpendicular to c_2 $\leftrightarrow$ r_1c_2
r_2 perpendicular to c_1 $\leftrightarrow$ r_2c_1
r_3 perpendicular to c_1 $\leftrightarrow$ r_3c_1

Write down the remaining such correspondences between the framework and its graph.

From the above discussion of the braced rectangular framework of Problem 2.2, including the results of Problem 2.4, we have shown that all of the conditions of Theorem 2.1 are satisfied for this framework and are expressible in terms of paths in its associated bipartite graph. The framework is rigid and its bipartite graph is connected.

Now consider the following braced rectangular framework, which we know is not rigid, and its associated bipartite graph.

We saw a deformation of it in Section 2.1.

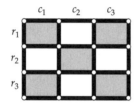 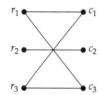

We can show that row r_1 must remain parallel to row r_3 (corresponding to path $r_1c_1r_3$ in the graph), and that column c_1 must remain parallel to column c_3 (corresponding to path $c_1r_1c_3$ in the graph). But we cannot show that row r_2 must remain parallel to row r_1, nor that row r_2 must remain parallel to row r_3, and correspondingly there are no paths in the bipartite graph from vertex r_2 to either vertex r_1 or vertex r_3. Nor can we show that column c_2 is parallel to either column c_1 or column c_3, and correspondingly there are no paths in the bipartite graph from vertex c_2 to either vertex c_1 or vertex c_3. We can make similar observations about the perpendicularity or otherwise of pairs of rows and columns and the existence or otherwise of corresponding paths in the bipartite graph.

We summarize our discussion in the following theorem.

Theorem 2.2

A braced rectangular framework is rigid if and only if its associated bipartite graph is connected.

Problem 2.5

By constructing the associated bipartite graph, determine whether each of the following braced rectangular frameworks is rigid.

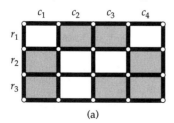 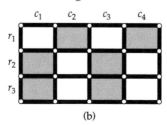

(a)   (b)

2.3 Minimum bracings

We have seen that the associated bipartite graph of a rigid braced rectangular framework is connected. If the graph has a cycle, then the removal of any edge of this cycle does not disconnect the graph. Hence the new graph (with one fewer edge) represents another rigid bracing of the framework. It follows that the framework with the corresponding brace removed is still rigid.

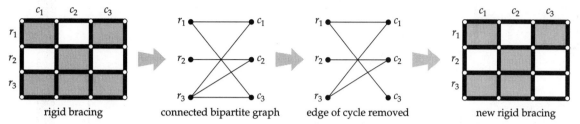

rigid bracing · connected bipartite graph · edge of cycle removed · new rigid bracing

If we repeat this procedure and continue to remove edges from cycles in the bipartite graph of a rigid framework, we eventually obtain a tree. Since in this process we destroy all the cycles and retain every vertex, this tree is a *spanning* tree of the bipartite graph.

We have removed the edge $r_3 c_3$ of the cycle $r_1 c_1 r_3 c_3 r_1$.

This leads to the following definitions and theorem.

Definitions

A rigid braced rectangular framework is **minimally braced** if no brace can be removed without destroying the rigidity; the corresponding bracing is a **minimum bracing**.

The connection between minimum bracings and spanning trees is summarized in the following theorem.

Theorem 2.3

A rigid braced rectangular framework is minimally braced if and only if its associated bipartite graph is a spanning tree.

The following familiar properties of trees give us a useful corollary of Theorem 2.3:

- a tree with n vertices has $n - 1$ edges;

- a tree does not contain a cycle.

Corollary

If the bipartite graph associated with a rigid braced rectangular framework has either of the following properties, then the corresponding bracing is not a minimum bracing:

- the graph has n vertices and more than $n - 1$ edges;

- the graph contains a cycle.

Problem 2.6

(a) By constructing the associated bipartite graph, show that the braced rectangular framework shown in the margin is minimally braced.

(b) Construct another minimum bracing for a 3×4 rectangular framework.

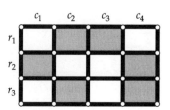

We can derive further minimum bracings from a given minimum bracing by considering various alterations to the bipartite graph which maintain the rigidity. If we insert a brace into an unbraced bay in a minimum bracing, then we create a cycle in the corresponding bipartite graph, and of course the new framework remains rigid.

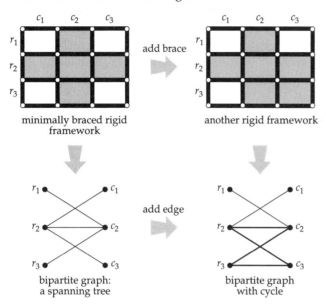

The new bipartite graph created by adding a brace to the above framework is no longer a spanning tree since it contains the cycle $r_2c_2r_3c_3r_2$. But if we now remove a *different* edge (that is, not the one we added) from this cycle, we obtain a spanning tree different from the earlier one. Since the cycle we created has length 4, we can remove any one of its four edges. This leads to four different spanning trees: the original tree and three others, which give rise to three new minimum bracings for the framework.

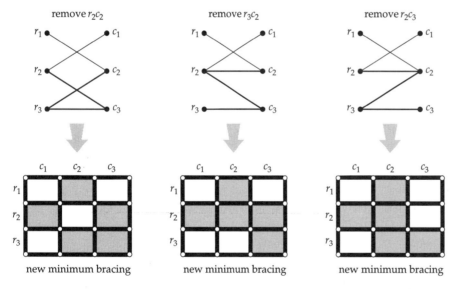

Finally, we can derive new rigid bracings from a given one in the following way. If we permute the rows (or columns) of a rigid braced rectangular framework, we do not change the parallelism or perpendicularity relationships that must be maintained between the rows and columns of links. Hence the bracing remains rigid. This is perhaps seen most clearly by observing what happens to the associated bipartite graph in the process.

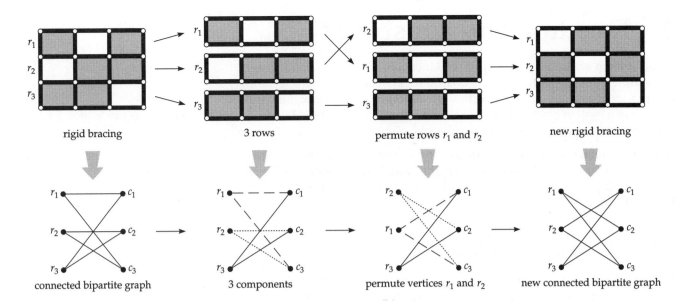

| rigid bracing | 3 rows | permute rows r_1 and r_2 | new rigid bracing |

| connected bipartite graph | 3 components | permute vertices r_1 and r_2 | new connected bipartite graph |

As you can see from the above figures, permuting the rows of the framework corresponds to moving the associated vertices in the bipartite graph without adding or deleting any of the edges. So, essentially, the vertices of the graph are just relabelled and the structure of the graph remains unchanged.

2.4 Computer activities

The computer activities for this section are described in the *Computer Activities Booklet*.

After studying this section, you should be able to:

- recognize a *braced rectangular framework* and give some examples;

- explain the terms *brace* and *bay*(i, j);

- write down the relationships between pairs of rows and/or columns of a braced rectangular framework that must hold if it is to be rigid;

- understand the *bipartite graph representation* of a braced rectangular framework and construct the appropriate graph for a given framework;

- understand the relationship between the rigidity of the braced rectangular framework and the connectedness of the associated bipartite graph;

- explain the terms *minimally braced* and *minimum bracing* and the relevance of spanning trees;

- construct other rigid braced rectangular frameworks from ones already known to be rigid.

3 Freedom and constraint

In this section we look at two important kinematic concepts — *freedom* and *constraint*. They provide us with quantitative measures of the general kinematic behaviour and properties of a system. We show how the *mobility* of a system can be determined from a combination of the freedoms and the constraints of its component links and joints.

We defined *mobility* in Section 1.4.

We restrict ourselves, in this section, to systems consisting of just *two* links. Then in Section 4 we look at more complicated systems consisting of several links and several joints, and derive mobility criteria for planar kinematic systems.

3.1 Kinematic freedom

The analysis of a kinematic system with only one (nullary) link is not very illuminating or interesting from a combinatorial point of view, and so we begin with a two-link system. For this, and for more complicated systems, we attach the reference coordinate frame to one of the links, which we refer to as the *fixed link*. All motions are then internal motions relative to this fixed link.

We defined *fixed link* in Section 1.4.

So, consider a system consisting of two links, in which one link is completely free in its movements relative to the other (that is, they are not connected together in any way). Both links are then nullary links. We shall see that we require *six independent coordinates* in order to describe the pose of one link with respect to the other (fixed) link at any instant in time. By this we mean that, in order to describe fully the position and orientation of one link relative to the other at a particular instant, we must be supplied with at least *six* quantities, such as the coordinates of points or the angles specifying directions of lines in the links. For example, consider how we might specify the pose of an aircraft relative to the ground (regarded as the fixed link). This is usually described in terms of three coordinates giving the position (such as *latitude*, *longitude* and *height* above sea-level), and a further three coordinates giving the orientation (such as *yaw*, *pitch* and *roll* angles), making a total of six coordinates.

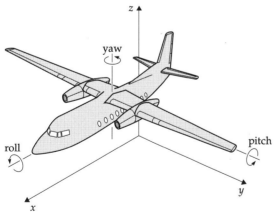

Definitions

The **freedoms of a link** are the independent quantities or coordinates that must be specified in order to describe the pose of the link with respect to some reference coordinate frame.

The **freedoms of a two-link kinematic system** are the independent quantities or coordinates that must be specified in order to describe the pose of one link of the system with respect to some reference coordinate frame fixed in the other link of the system.

A system of two nullary links (the aircraft and the ground, for instance) therefore has *six* freedoms.

Consider a two-link kinematic system consisting of the Earth and an orbiting octahedral spacecraft. If we treat the Earth as the fixed link, and attach a reference Cartesian coordinate frame to its centre, then we can define the pose of the moving spacecraft relative to this frame.

We choose an *octahedron* because this has six vertices, and we shall eventually need to identify six points on it where we can restrict its freedom. In the terminology of Section 1, the octahedral spacecraft is potentially a 6-ary link.

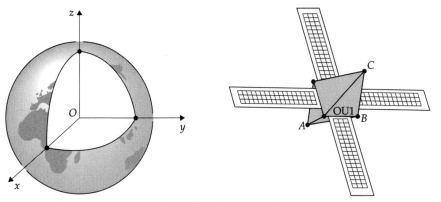

the Earth and a spacecraft

Remember that we must take account of both the *position* and the *orientation* of the octahedral spacecraft in order to specify its pose completely. This can be done in terms of the locations of the vertices of the octahedron and/or the locations of some of the edges joining these vertices.

We begin by choosing a point, such as the vertex A, of the spacecraft. We can describe the position of A by its three Cartesian coordinates x_A, y_A, z_A, for example. But these three coordinates do not completely describe the pose of the spacecraft, since it is still free to pivot about A.

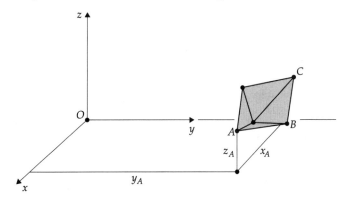

We next determine the direction of the edge AB joining A to an adjacent vertex B. This requires two further coordinates, such as the angles θ_B and ϕ_B which are essentially the latitude and longitude of B if it were on the surface of a sphere centred at A. The five coordinates x_A, y_A, z_A, θ_B, ϕ_B locate the position of the edge AB, but the spacecraft is still free to rotate about this edge.

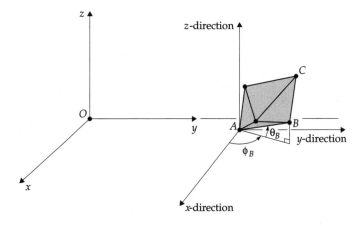

39

In order to specify the spacecraft's pose completely, we need one more coordinate, making six in all. The most obvious one to choose is an angle ψ_C which describes the angular position (the *azimuth*) of a third corner C with respect to the x-direction (say), as the craft swivels about the edge AB.

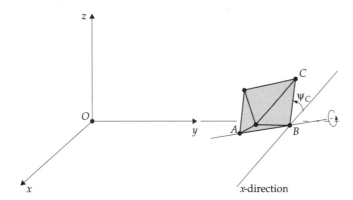

The pose of the spacecraft with respect to the Earth is now completely defined when the six coordinates $x_A, y_A, z_A, \theta_B, \phi_B$ and ψ_C are given specific values.

Problem 3.1

(a) Suggest an alternative set of six coordinates to describe the pose of the spacecraft.

(b) Why can we not take the x, y and z coordinates of *two* points A and B, so that our six coordinates would be $x_A, y_A, z_A, x_B, y_B, z_B$?

In our discussion of the pose, three of the freedoms were concerned with *position* (x_A, y_A and z_A) and three were concerned with *orientation* (θ_B, ϕ_B and ψ_C). However, as you can see from the solution to the above problem, this division into two types of freedom is purely accidental and *any* six *independent* coordinates can be used, regardless of their nature.

Notice that we needed *three distinct non-collinear points* — the vertices A, B and C — on the spacecraft in order to define the coordinates x_A, y_A, z_A, θ_B, ϕ_B, ψ_C. Similarly, in our solution to Problem 3.1(a) we required these three points to define the alternative set of coordinates x_A, y_A, y_B, z_B, z_C, x_C. This is true in general.

Theorem 3.1

We must always consider at least three points when describing the freedoms of a link in space. Furthermore, once we have specified the positions of any three distinct non-collinear points of a link, then we have specified the pose of the whole link.

The reason why we need to consider at least three points in order to describe the pose of a link in space is quite simple. If we used just two points, A and B, on the link, then they must always remain a fixed distance apart (since a link is a *rigid* body). Thus they cannot move completely independently of each other. This dependence is expressed mathematically by an equation relating the six coordinates $x_A, y_A, z_A, x_B,$ y_B, z_B — in this case it is the three-dimensional version of the distance formula attributed to Pythagoras,

$$(x_B - x_A)^2 + (y_B - y_A)^2 + (z_B - z_A)^2 = d^2$$

where d is the fixed distance between A and B.

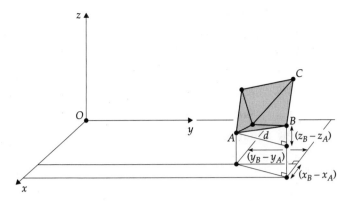

We have six coordinates and one equation of dependence; this means that only $6 - 1 = 5$ coordinates are independent, so this set of coordinates based on just two points is unsatisfactory.

3.2 Kinematic constraint

Let us now return to the situation where we have chosen six *independent* coordinates to describe the freedoms of the two-link kinematic system comprising an octahedral spacecraft orbiting the Earth. If we restrict the motion of the spacecraft in some way, it no longer possesses all six freedoms. One or more of the freedoms have been constrained and we say that we have imposed *kinematic constraints*.

Definition

A **constraint** on a link is a geometric restriction that removes *one* of its freedoms.

You have already seen several examples of kinematic constraint, since a kinematic pair (binary joint) imposes one or more constraints on the relative freedom of the two links it connects. A further example of a kinematic constraint is provided by restricting the pose of the spacecraft so that one of its vertices, A say, always lies on a certain plane, say the equatorial plane of the Earth.

This is not an unreasonable restriction, since any orbiting spacecraft is constrained by gravity to move along an elliptical orbit, and an ellipse is a plane curve.

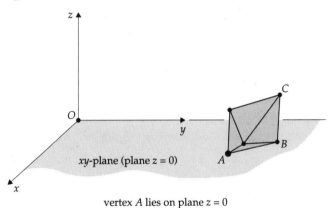

vertex A lies on plane $z = 0$

We can describe such a constraint mathematically by a single equation involving one or more of the six coordinates. For example, if A cannot move out of the equatorial plane, then the coordinate z_A must be constant, and in this case equal to zero. The *constraint equation* is therefore very simple and is

$$z_A = 0 \tag{3.1}$$

We have introduced the term *mobility* to describe the net number of independent coordinates required to specify the pose of every link of a system, with respect to a coordinate frame on the fixed link, when the constraints are taken into account. So in this case, the mobility of the spacecraft relative to the Earth is 5, since only five (that is, 6 – 1) coordinates are now required to describe its pose. These coordinates might be x_A, y_A, θ_B, ϕ_B and ψ_C. The single constraint equation $z_A = 0$ has reduced the number of freedoms by 1.

Problem 3.2

For each of the following surfaces, draw rough sketches to illustrate the constraint on the spacecraft, and state the constraint equation, if the vertex A must lie on the surface:

(a) the plane $x + y + z = 4$

(b) the sphere $x^2 + y^2 + z^2 = 9$

(c) the cylinder $x^2 + y^2 = 4$

(d) the cone $x^2 + y^2 = z^2$

We can see from our solution to Problem 3.2 that, if we know the equation of the surface on which a point of the spacecraft is constrained, then it is easy to obtain the constraint equation. This solution also illustrates several other important facts, which we list here for reference:

- a constraint equation usually involves several, if not all, of the six coordinates specifying the freedoms of a two-link kinematic system, and *must* involve at least one coordinate;

  For example, equation 3.1 involves just one coordinate.

- if a constraint equation involves only one coordinate, then we can restate the original problem so that this coordinate is excluded from the outset, and we begin with the non-fixed link having only five freedoms;

- a constraint equation can be linear (case (a)), quadratic (cases (b), (c) and (d)), any other type of polynomial (cubic, quartic, etc.), trigonometric, and so on — the *type* of equation does not matter.

So far we have considered only a single constraint, specified by only one constraint equation. In general, however, we may have more than one constraint equation. If we return to the spacecraft and assume that it is in orbit around the equator, then we see that a more accurate description of the constraint keeping it 'on course' actually involves two surfaces and not just one. The spacecraft is not free to move over the *whole* surface of the equatorial plane, but is restricted by the laws of celestial mechanics to travel along a definite curve (a conic section, such as an ellipse or circle) — its orbit. This curve can be thought of as the curve of intersection between two surfaces. For a circular orbit, for example, the surfaces could be a sphere and a plane, or a cylinder and a plane. In either case, the vertex A of the spacecraft is constrained to lie on both surfaces at the same time, so that we have *two* constraint equations.

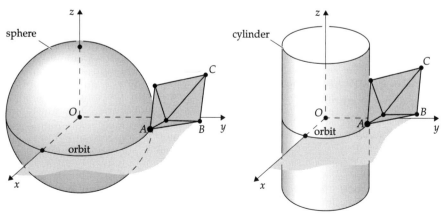

vertex A traces circle on sphere   vertex A traces circle on cylinder

Problem 3.3

What is the mobility of the spacecraft system when these *two* constraints are taken into account?

We can continue in this manner and add a third constraint (specified by a third constraint equation) on the vertex A, so that the spacecraft is fixed in space at the point of intersection of *three* surfaces. The spacecraft can then only pivot about this point, and the mobility of the system is reduced to 3, since the spacecraft's pose is now adequately described by θ_B, ϕ_B and ψ_C alone.

However, we cannot add a fourth constraint to the vertex A, since it is already immobilized in space, although the rest of the spacecraft is not. In order to proceed any further, we must constrain *another* vertex, say B. We find that we can impose *at most two* such constraints on B, and then it too is immobilized in space. A sixth constraint must be imposed on a third vertex C; and, if this is done, the whole spacecraft is completely fixed in space. The system's mobility is then 0.

Each time we impose a new *independent* constraint on one or more points, we introduce another *independent* constraint equation and, by the definition of a constraint, we reduce the mobility of our system by 1. This pattern is summarized by the following simple equation:

$$M = F - C \qquad\qquad (3.2)$$

where M is the mobility, F is the number of freedoms and C is the number of constraints. This fundamental equation is quite general.

Note that constraints may be imposed in many different ways. For example, we immobilized the spacecraft by distributing the necessary *six* constraints among *three* vertices A, B and C so that we had three constraints on A, two on B and one on C. However, there are several other ways of distributing constraints among the six vertices of the octahedral spacecraft, since the number 6 may be partitioned in many different ways. In fact, there are eleven different partitions of 6, namely:

(a) 6

(b) $5 + 1$

(c) $4 + 2$

(d) $4 + 1 + 1$

(e) $3 + 3$

(f) $3 + 2 + 1$

(g) $3 + 1 + 1 + 1$

(h) $2 + 2 + 2$

(i) $2 + 2 + 1 + 1$

(j) $2 + 1 + 1 + 1 + 1$

(k) $1 + 1 + 1 + 1 + 1 + 1$

Problem 3.4

Only six of the eleven partitions of 6 can be utilized for constraining the spacecraft. Identify these, and explain why each of the remaining five partitions of 6 is inadmissible.

Earlier we used partition (f) (3 constraints on A, 2 on B, 1 on C) to immobilize the spacecraft. Later, we shall also use partition (h) (2 constraints on A, 2 on B, 2 on C) and partition (k) (one constraint on each of six different points) as they are rather interesting. In fact, partition (k) provides a justification for choosing an octahedron as the shape of our spacecraft, because an octahedron has six vertices and these provide six convenient points at which to impose a constraint.

Problem 3.5

Consider a flat object, such as a thin triangular slab, whose motion is restricted to two-dimensional space — the plane in which it lies.

(a) How many freedoms does the slab have initially?

(b) How many constraints are required to immobilize the slab?

(c) How many points on the slab need to be fixed in order to immobilize it?

(d) List the possible ways of immobilizing the slab.

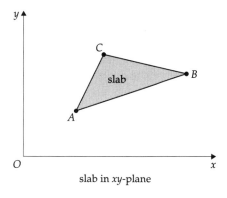

slab in xy-plane

3.3 Connectivity of kinematic pairs

In the previous subsection we saw how we can impose a number of constraints on a link in several different ways, by partitioning the number of constraints and distributing the constraints correspondingly among distinct points of the link. We assumed that the constraints are *independent*. However, this does not mean that each constraint surface is necessarily distinct. We may often impose the *same* constraint on two *distinct* points, and we must then count this as *two* constraints. For example, suppose that the constraint surface is a plane with the equation

$x + y + z = 4$

If we constrain the two vertices A and B (say) of our spacecraft so that they both always lie on the plane, then we have effectively imposed *two* constraints, because we now have *two* independent constraint equations that must be satisfied simultaneously — namely

$x_A + y_A + z_A = 4$ and $x_B + y_B + z_B = 4$

In this case, the edge AB always lies in the constraint plane as the octahedron moves, and the spacecraft now has only four (6 − 2) freedoms, instead of the original six. In other words, its mobility has been reduced to 4. We can verify this directly by listing the four coordinates now required to specify the pose of the octahedron — two coordinates for locating the position of A on the constraint plane, one coordinate for specifying the angle describing the orientation of the edge AB on the plane, and one further coordinate for specifying the rotation angle of the octahedron about the edge AB.

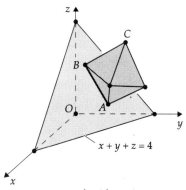

spacecraft with vertices A, B on plane

If three non-collinear points of the octahedron are constrained to lie on the above constraint plane, then three constraints are imposed, and the spacecraft has only three (6 − 3) freedoms remaining. These consist of two translational freedoms in two independent directions on the plane, together with a single rotational freedom about a line perpendicular to the plane.

In general, a maximum of three non-collinear points of a link may be constrained in this way by a plane surface, and the link thereby loses at most three freedoms. The remaining three freedoms (from the original six) are not affected by restricting a fourth, fifth or sixth point to lie on the plane. We can demonstrate this by thinking of the link as a kitchen stool (instead of as an octahedrally shaped spacecraft) and noting that a four-legged stool has the same mobility — that is, 3 — as a three-legged stool, provided that the legs are kept in contact with the floor, which is flat.

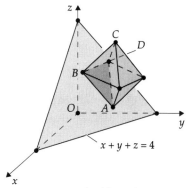

spacecraft with vertices A, B, D on plane

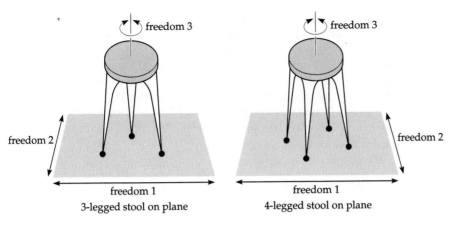

freedom 3

freedom 3

freedom 2

freedom 2

freedom 1

freedom 1

3-legged stool on plane

4-legged stool on plane

We idealize the stool so that its feet are single points rather than contact patches.

Problem 3.6

Consider the following three surfaces:

(a) a cylinder;

(b) a sphere;

(c) a cone.

In each case, state the maximum number of non-collinear points of a link that may be constrained to lie on the surface, such that if any further points are so constrained then the link does not lose any more freedoms.

Hint Think of a stool whose feet represent the points, and determine the required number of feet.

We have now seen how various constraints are imposed on one or more points of a link by restricting the points to lie on one or more constraint surfaces. In this way, the number of freedoms of the link with respect to a fixed coordinate frame is reduced from 6 (when there are no constraints) to $6 - k$ (when there are k independent constraints). How can we apply these results to explain the freedoms of the links in a two-link kinematic system, relative to each other, when the single binary joint is one of the six Reuleaux pairs shown in Table 1.7? In each case, the two links constrain each other by having common surfaces in contact rather than discrete points, as already explained. However, we now show that this surface contact can be viewed in terms of discrete point contact, although the former description is more convenient for our subsequent developments. We illustrate the technique for such a system involving a revolute pair.

A two-link system joined by a revolute pair has only a single relative freedom — a rotational freedom. If the two links involved were not connected, each would have six freedoms relative to a coordinate frame fixed in the other. The revolute pair must therefore impose five constraints. How does it do this in terms of constraint surfaces constraining discrete points?

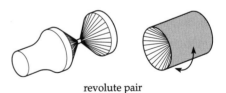

revolute pair

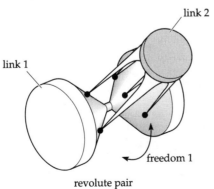

link 2

link 1

freedom 1

revolute pair

45

If we imagine that we remove material from one of the links, and continue to do so without destroying or changing the rotational freedom, then we find that we must stop when we reach the stage of having only five points in contact with the second link. This five-point contact represents the minimum possible 'skeleton' connection that can simulate the revolute pair. Each of the five points is constrained by the conical surface, so that the required five constraints are thereby imposed. This arrangement allows only rotation about the cone's axis to take place, and prevents sliding along the direction of the axis (or any other motion) from occurring, since such a motion would move at least one of the points off the surface.

We illustrate the situation in Table 3.1, together with the 'skeletal' arrangements for two-link systems involving the other five Reuleaux pairs.

Recall that the contact surface for a revolute pair is a cone, and we found in Problem 3.6(c) that a maximum of five points of a link can be constrained by a cone.

Table 3.1 Connectivity of two-link systems involving Reuleaux pairs.

Reuleaux pair	'skeleton' form	points of contact (number of connections)	freedoms (connectivity)
revolute pair		5	1
prismatic pair		5	1
screw pair		5	1
cylindric pair		4	2
planar pair		3	3
spherical pair		3	3

The 'skeleton' forms can be deduced from our knowledge of the Reuleaux pairs and from our solution to Problem 3.6.

Definition

The **connectivity** of a kinematic pair, considered as the only joint in a two-link kinematic system, is the number of freedoms of one of the links with respect to the other.

In using the notion of *connectivity* here, we must avoid confusion with the concept of *connectedness*. In fact the term 'connectivity' in the context of kinematics is something of a misnomer. For example, the connectivity of a revolute pair is only 1, whereas we have just shown that a revolute pair is simulated by *five* 'connections'; and a cylindric pair has a higher connectivity (2), but is simulated by fewer connections (4). We have the following relationship for a Reuleaux pair:

$$\text{connectivity} = 6 - \text{number of connections.} \qquad (3.3)$$

This is just another way of stating the basic relationship between mobility, freedoms and constraints, for a system of two links, one of which is considered to be the fixed link. In other words, the connectivity of a Reuleaux pair is simply the mobility of the two-link system.

Compare equations 3.2 and 3.3.

Henceforth we shall not use the skeletal form of the Reuleaux pairs, and we shall treat them as indivisible building blocks. However, we shall continue to refer to the number of freedoms of their corresponding two-link systems (that is, the mobility of the two-link systems) as their connectivity.

In our discussions so far, we have not considered how we can impose our constraints in practice. We might think that we could just make a smooth surface (a sphere, a plane, and so on) and place the appropriate point of the rigid body on this surface. However, the difficulty lies in how we can maintain contact throughout the motion, since the body will tend to lift off the surface. In practice, we usually impose the constraints in a different way. Since a common type of constraint is that in which a point is forced to lie on the surface of a sphere, we illustrate the procedure with this example.

Let us return to our spacecraft and constrain the vertex A to move on the surface of a sphere described by the equation $x^2 + y^2 + z^2 = 4$. This sphere has radius 2, and so A is always at a distance 2 from the centre O of the sphere. If, instead, we join A to the centre O with a link of length 2, and connect this link to the reference frame at O and to the spacecraft at A by means of two *spherical* pairs, then effectively we have imposed our constraint by introducing an extra binary link into the system together with two spherical pairs.

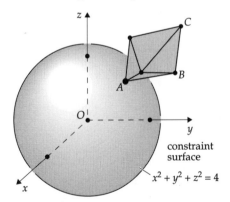

vertex A lies on a sphere of radius 2

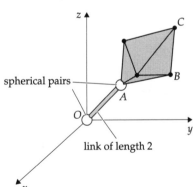

binary link from A to O of length 2

Further sphere constraints can be added in the same way. We illustrate two important examples of how we can immobilize the spacecraft with six such constraints by specifying that the vertices A, B, C, etc. lie on various spheres centred at O_1, O_2, O_3, etc.

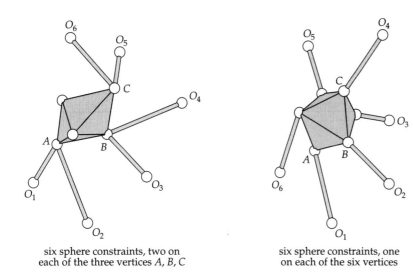

six sphere constraints, two on
each of the three vertices A, B, C

six sphere constraints, one
on each of the six vertices

We can adopt the same sort of procedure to constrain the triangular slab of Problem 3.5. In this case, we consider a typical two-dimensional constraint — in which the vertex A lies on a circle — and we use a binary link to impose the constraint. It is attached at O and A by *revolute* pairs.

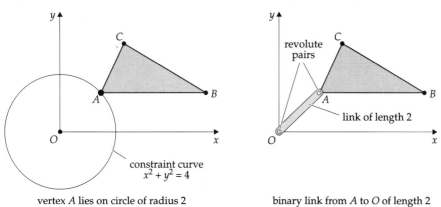

vertex A lies on circle of radius 2

binary link from A to O of length 2

We can completely immobilize the triangular slab with three such constraints, one on each vertex.

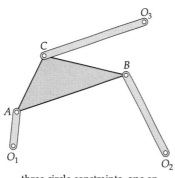

three circle constraints, one on
each of the three vertices

As you can see from these examples, essentially we have introduced extra binary links into our system in order to constrain the single free link (the spacecraft or triangular slab). This is usually what happens in any real kinematic system, as you will see in Section 5 with the flight simulator. It is far more realistic, therefore, to consider the mobility of systems containing more than two links. This we do in Section 4.

After studying this section, you should be able to:

- explain the terms *freedom, constraint, mobility, constraint surface* and *constraint equation,* with reference to kinematic systems, and remember how they are interrelated;

- suggest appropriate values for these quantities for some simple kinematic systems;

- understand the meaning of the *connectivity* of a Reuleaux pair;

- understand the idea of 'simulating' constraint surfaces by introducing extra links and joints into the system.

4 Planar kinematic systems

In this section we discuss the special case of *planar* kinematic systems, and particularly those with mobilities 0 and 1. In Section 1, we defined planar systems as those in which all the links move only in parallel planes. We saw that only three of the six Reuleaux pairs can occur in planar systems: the *revolute pair*, the *prismatic pair*, and the *planar pair*. When revolute pairs are present in the system, all the revolute axes must be parallel, in a direction perpendicular to the planes in which motion takes place. If the system has prismatic pairs, then these must be arranged so that their axes are parallel to the planes of motion, although they need not necessarily be parallel to each other. Finally, if the planar system has any planar pairs, then the plane of contact between the two links joined by each such pair must be parallel to the planes of motion.

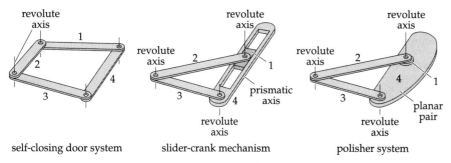

self-closing door system   slider-crank mechanism   polisher system

A typical planar kinematic system containing only revolute pairs is the self-closing door we considered in Section 1. This system consists essentially of four binary links connected by four revolute pairs. A common planar kinematic system containing a prismatic pair is the slider-crank mechanism used as the basis of many engine or pump designs. This has four binary links connected by three revolute pairs and one prismatic pair, making four pairs in all. A system containing a planar pair is often used as the basis of a mechanism for polishing a flat surface, and a similar device is used for grinding and polishing lenses for telescopes, despite the fact that the lens surface is not flat. We illustrate an example of such a system, which consists of four binary links connected by three revolute pairs and one planar pair, again making four pairs in all.

4.1 Systems with only revolute pairs

For the remainder of Section 4, we restrict our attention to those planar kinematic systems containing only revolute pairs, since they are widely used and are relatively straightforward to analyse. We can draw a schematic diagram representing the 'plan view' of such a system, and the revolute pairs then appear as small circles or *points*. If the system contains only binary links, then this schematic diagram can be used as the basis for a *direct* graph representation. However, we wish to consider more general systems containing ternary, quaternary, ... links, so the *interchange* graph representation is usually more useful.

We have already seen an example of a complicated planar system containing only revolute pairs: the set-square mechanism discussed in Section 1. This consists of eight links: five binary links (1, 2, 3, 5 and 6), one ternary link (7), one quaternary link (4), and one unary link (8). All the joints are revolute pairs, and, since these are binary joints, we can use the interchange graph representation to describe the system.

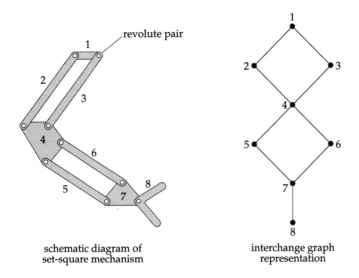

schematic diagram of
set-square mechanism

interchange graph
representation

The interchange graph for this mechanism resembles two single-loop four-link systems joined together at a common vertex (vertex 4). In fact, the physical system is structured in just this way. We may envisage constructing it from two single-loop four-link systems by combining a binary link from each system into a single quaternary link.

Here we ignore the unary link — the set-square itself.

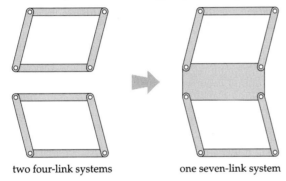

two four-link systems   one seven-link system

An alternative method of construction is to start with one single-loop four-link system, then to expand one of the binary links into a potential quaternary link by replacing it with a quadrilateral, and finally to attach a system of three links across the corners of the quadrilateral.

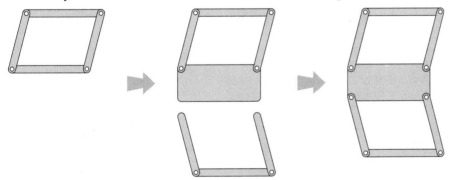

This second method is usually better for constructing new systems from existing ones, in the sense that it is easier to formulate and apply the rules of construction in this case. It is also a more straightforward procedure in terms of graph operations, since all we have to do is to introduce new vertices, together with appropriate edges, in the corresponding interchange graph.

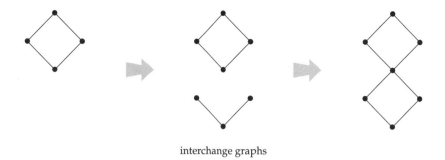

interchange graphs

4.2 Factors affecting mobility

It can be quite difficult to determine the mobility of a given planar kinematic system with only revolute pairs. The system may have many links interconnected by many revolute pairs (each of which imposes constraints on the component links), and the whole system is arranged in a particular geometric configuration, which itself may have an effect on the mobility.

In order to demonstrate that the geometry of the system can be important, consider a single-loop four-link planar system.

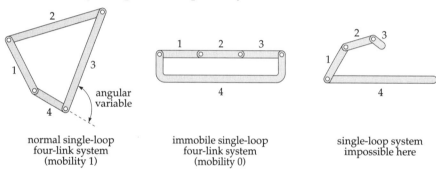

| normal single-loop four-link system (mobility 1) | immobile single-loop four-link system (mobility 0) | single-loop system impossible here |

We have already shown that, in general, this has mobility 1, because a single angle is sufficient to specify the relative poses of the component links. However, if the 'lengths' of three of the links add up to the 'length' of the fourth link, then the system cannot move (that is, it has mobility 0), despite the fact that we have not altered the interconnections in any way. In fact, we can choose the link lengths so that we cannot even connect the system together in the way required.

The geometry of a system sometimes *reduces* its mobility; but it is more common for the geometry to introduce more freedom and hence to *increase* the mobility. A typical situation where the mobility can be greater than expected occurs in a planar five-link system. If we construct such a system using two ternary links, three binary links and only revolute pairs, we obtain an arrangement such as that below in (a).

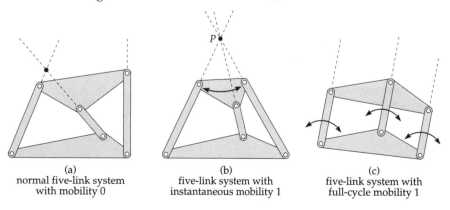

| (a) normal five-link system with mobility 0 | (b) five-link system with instantaneous mobility 1 | (c) five-link system with full-cycle mobility 1 |

For an arbitrary choice of link lengths or link shapes, as in (a), the system is an immobile structure, and thus has mobility 0. If, however, the dimensions of the links are such that the three imaginary straight lines joining the two revolute pairs on each binary link all intersect in a single point P, as in (b), then the components of the system can move, relative to one another, through a small range on either side of their 'equilibrium' position. In practice, we usually obtain quite a reasonable range of motion about the equilibrium position because of a certain amount of 'stretch' and 'give' in the physical components.

No material is completely rigid, and so it deforms slightly to accommodate small stresses.

Such *extraneous mobility* must be avoided in the design of framework structures, where such movements might lead to the catastrophic collapse of the supposedly immobile structure. Indeed, even greater care must be taken to avoid the more serious introduction of *full-cycle mobility* into the structure, as in (c), where the three lines remain parallel throughout the large range of motion.

Before we leave the subject of geometric complications in the determination of mobility, we briefly mention *limit positions*. These are positions where, during the motion of a system, some of the component links of the system momentarily come to rest and then reverse their direction of motion. For example, link 1 in the diagram below has reached the limit of its anticlockwise rotation about the revolute pair A, and it is now free to rotate only in the clockwise sense. If, in addition, the clockwise motion of the system is constrained by other links in a limit position, then it would be effectively immobilized, despite the fact that the system is supposed to have mobility 1.

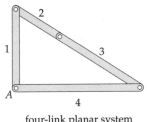

four-link planar system
at a limit position

We have now discussed some of the problems associated with the determination of the mobility of a kinematic system. We have seen that a knowledge and understanding of the detailed geometry is required (in addition to a specification of the number, type and arrangement of links and joints) in order to obtain the true mobility with confidence. However, from now on, we ignore the purely geometrical aspects, and proceed to derive various formulae for determining a quantity that we continue to refer to as the *mobility*, despite the fact that for certain geometric configurations these formulae fail to give the true value for the mobility. We refer to these formulae collectively as **mobility criteria**, and they are expressed purely in terms of combinatorial quantities such as the number of links, the number of revolute pairs, and so on.

4.3   Planar mobility criteria

Recall from Section 3 the relationship between the mobility M, the number of freedoms F, and the number of constraints C in a two-link kinematic system:

$$M = F - C \tag{4.1}$$

Equation 3.2 of Section 3.

We shall expand the terms in this equation so that it can be applied to planar systems containing many links and joints. The process is straightforward and merely involves the substitution of appropriate expressions for F and C.

We can easily obtain the expression for F by noting that a link which is free to move in a plane has only *three* freedoms instead of the six it would have in space.

In a planar system containing n nullary links, with one link considered to be the fixed link, we have $n - 1$ links moving relative to a coordinate frame attached to the fixed link. So there are $3(n - 1)$ relative freedoms altogether, since the links are unconnected. Thus

$$F = 3(n - 1)$$

Since there are no joints in this system (so that $C = 0$), this is also the value of M.

Now, in general, each revolute pair that we introduce into a *spatial* system imposes *five* constraints, because its connectivity is 1. However, when all the revolute axes are parallel, the sole function of three of these five constraints is to constrain the total motion of the system to be planar, rather than spatial. These three constraints are said to impose a **general constraint** on the system, and we have already taken account of this general constraint by allowing each link to have only the three freedoms allowed by a planar system, instead of the six allowed by a spatial system. Each revolute pair in a planar system therefore imposes only *two* $(5 - 3)$ constraints on the freedom of the planar motion. So if there are j revolute pairs in the system, we have thus introduced $2j$ constraints, and our expression for C is

$$C = 2j$$

Substituting the above expressions for F and C into equation 4.1, we obtain a simple formula for determining the mobility of a planar kinematic system containing only revolute pairs, provided that we ignore any geometrical complications.

Theorem 4.1: first mobility criterion

In a planar kinematic system containing only revolute pairs, the mobility M of the system relative to one of the links (considered as the fixed link) is given by the mobility criterion

$$M = 3(n - 1) - 2j$$

where n is the total number of links (regardless of their multiplicity) and j is the number of revolute pairs.

The value obtained for M from the mobility criterion in Theorem 4.1 is always an integer. There are three possible cases.

If M is *positive* and if there are no geometrical complications, then the links of the planar system can have relative motion and the mobility of the joints must be controlled in order to control the motion of the system. The system is said to be **mobile**.

If $M = 0$ and if there are no geometrical complications, then the links of the system have no relative motion and the system is said to be **immobile**.

If M is *negative* and if there are no geometrical complications, then the links of the system again have no relative motion. But, in addition, some of the links and joints can be removed without allowing the remaining links to have any relative motion. The system is said to be **overconstrained**.

Example 4.1

The mobility of the self-closing door discussed in Section 1.4 is 1, since only one quantity — the angle the door makes with its frame — is required to specify the pose of every link of the system with respect to the fixed link — the door frame.

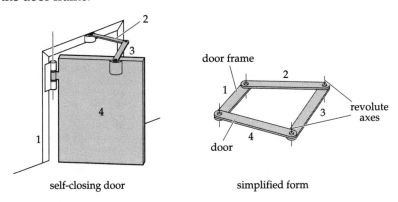

self-closing door   simplified form

We can verify this mobility value by using the mobility criterion in Theorem 4.1. There are four binary links and four binary joints in the system, so $n = 4$ and $j = 4$. Hence, we have

$$M = 3(4 - 1) - 2 \times 4$$

$$= 9 - 8$$

$$= 1$$

as expected. ∎

Problem 4.1

Use Theorem 4.1 to determine the mobility of each of the following planar systems with four links and only revolute pairs.

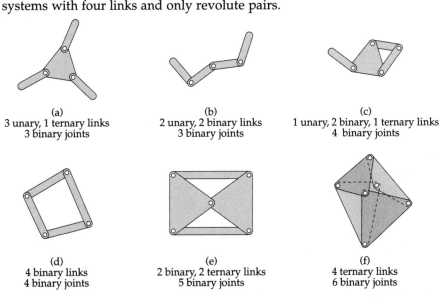

(a)	(b)	(c)
3 unary, 1 ternary links	2 unary, 2 binary links	1 unary, 2 binary, 1 ternary links
3 binary joints	3 binary joints	4 binary joints

(d)	(e)	(f)
4 binary links	2 binary, 2 ternary links	4 ternary links
4 binary joints	5 binary joints	6 binary joints

These are the systems shown in Table 1.3, with system (f) redrawn as a planar system.

The mobility criterion in Theorem 4.1 is often used in practice. However, it takes no account of the various types of link explicitly, so that there are no separate terms for the number of binary links, the number of ternary links, and so on. Furthermore, we frequently encounter multiple joints, and so it is useful to have a mobility criterion that is expressed in terms of the number of joints (regardless of their multiplicity), rather than just in terms of revolute pairs. Both of these requirements can be satisfied together in a single, more general, mobility criterion, which we now derive for a system containing a total of n links and g joints.

We continue to consider only revolute pairs, so that each r-ary joint should be regarded as a combination of $r - 1$ revolute pairs.

We begin by noting that if there are n_0 nullary links, n_1 unary links, n_2 binary links, n_3 ternary links, and, in general, n_r r-ary links, and if the system has no links with multiplicity greater than s, then all of these terms must add up to the total number of links n. We thus have the following equation:

$$n_0 + n_1 + n_2 + n_3 + \cdots + n_r + \cdots + n_s = n \qquad (0 \le r \le s) \qquad (4.2)$$

We can give a similar expression for the joints. If there are j_2 binary joints, j_3 ternary joints, j_4 quaternary joints, and, in general, j_r r-ary joints, and if the system has no joints with multiplicity greater than t, then all of these terms must add up to the total number of joints g. We thus have the following equation:

$$j_2 + j_3 + j_4 + \cdots + j_r + \cdots + j_t = g \qquad (2 \le r \le t) \qquad (4.3)$$

See Section 1.1.

We now replace each multiple joint by the appropriate number of binary joints (revolute pairs). Note that each r-ary joint represents $r - 1$ revolute pairs, so that j_r r-ary joints represent $(r - 1)j_r$ pairs. This enables us to rewrite equation 4.3 in the form

$$j_2 + 2j_3 + 3j_4 + \cdots + (r - 1)j_r + \cdots + (t - 1)j_t = j \qquad (4.4)$$

where j is the total number of revolute pairs in the system after the multiple joints have been expanded.

Finally, by counting the connection points of all the links in a system in two different ways, we can derive the following equation:

$$n_1 + 2n_2 + \cdots + rn_r + \cdots + sn_s = 2j_2 + 3j_3 + \cdots + rj_r + \cdots + tj_t \qquad (4.5)$$

A *connection point* of a link is a point at which the link is joined to another link.

The left-hand side of this equation is obtained by noting that each binary link has 2 connection points, each ternary link has 3 connection points, and in general each r-ary link has r connection points. The right-hand side is obtained in a similar manner by noting that two connection points meet at each binary joint, and in general that r connection points meet at each r-ary joint.

Adding equations 4.3 and 4.4, we obtain

$$2j_2 + 3j_3 + 4j_4 + \cdots + rj_r + \cdots + tj_t = g + j$$

and from this and equation 4.5 we have

$$n_1 + 2n_2 + 3n_3 + \cdots + rn_r + \cdots + sn_s = g + j \qquad (4.6)$$

The original mobility criterion, in Theorem 4.1, can now be rewritten in terms of $n_0, n_1, n_2, \ldots, n_r, \ldots, n_s$ and g by substituting for n from equation 4.2 and for j from equation 4.6. We thus obtain the following expression:

$$M = 3(n - 1) - 2j$$
$$= 3(n_0 + n_1 + n_2 + n_3 + n_4 + \cdots + n_r + \cdots + n_s - 1)$$
$$- 2(n_1 + 2n_2 + 3n_3 + 4n_4 + \cdots + rn_r + \cdots + sn_s - g)$$

After rearranging and simplifying the right-hand side of this equation, we finally derive the following mobility criterion.

Theorem 4.2: second mobility criterion

In a planar kinematic system containing links and joints of various multiplicities, where each r-ary joint is equivalent to $r - 1$ *revolute* pairs, the mobility M of the system relative to one of the links (the fixed link) is given by the mobility criterion

$$M = 2g + 3n_0 + n_1 - ((2r - 3)n_r + \cdots + (2s - 3)n_s) - 3,$$

where g is the total number of joints (regardless of their multiplicity), no link has multiplicity greater than s, and n_r is the number of r-ary links ($2 \le r \le s$).

Example 4.2

If there are no special geometric relationships among the link lengths or link shapes of the five-link system discussed in Section 4.2, the mobility of the system is 0, so it is immobile. Hence, *no* quantities are needed to specify the pose of every link of the system with respect to the fixed link.

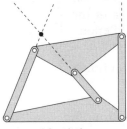

normal five-link system
with mobility 0

We can now verify this mobility value by using the mobility criterion in Theorem 4.2. There are two ternary links, three binary links and six binary joints (revolute pairs) in the system, so $n_3 = 2$, $n_2 = 3$ and $g = 6$. Also $n_r = 0$ for all values of r other than $r = 2$ or $r = 3$. Hence, we have

Remember that, in this section, the only binary joints considered are revolute pairs.

$$M = 2g - n_2 - 3n_3 - 3$$

$$= (2 \times 6) - 3 - (3 \times 2) - 3$$

$$= 12 - 3 - 6 - 3$$

$$= 0$$

as expected. ∎

Problem 4.2

Use Theorem 4.2 to determine the mobility of the eight-link planar kinematic system involving only revolute pairs shown in the margin.

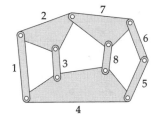

As was the case with Theorem 4.1, the value obtained for M from the mobility criterion in Theorem 4.2 is always an integer. Again, M can be positive, zero or negative, and, as before, the planar kinematic system is respectively mobile, immobile or overconstrained.

The two forms of the mobility criterion given by Theorems 4.1 and 4.2 are equivalent. Which version we use for a particular system is a matter of convenience. If all the joints are binary joints, it is simpler to use Theorem 4.1, where j is the number of such joints and n is the total number of links, regardless of multiplicity. Alternatively, if the system has multiple joints, it is better to use Theorem 4.2, where g is the total number of joints, regardless of multiplicity, and the various types of link (binary, ternary, etc.) must be counted separately as n_2, n_3, and so on.

Problem 4.3

Using the direct graph representation, derive the mobility criterion in Theorem 4.2 more directly for a system with only binary links, but with multiple joints.

Hint In this case, there are no links with multiplicity other than 2, so that $n_r = 0$ for all values of r other than $r = 2$. Consider the freedoms of the vertices (as points in the plane) and the constraints imposed by the edges.

As you can see from our solution to Problem 4.3, we can use Theorem 4.2 to discuss the *mobility of a graph* (with g vertices and n_2 edges). We consider the graph to be a direct representation of a planar system containing only binary links. This allows us to rewrite the mobility criterion of Theorem 4.2 in a form applicable to graphs.

Theorem 4.3: mobility criterion for graphs

In a simple connected graph whose vertices all have degree greater than or equal to 2, and which therefore can be considered to be the direct graph of a planar kinematic system containing multiple joints (each of which can be regarded as a combination of revolute pairs) and only binary links, the mobility M of the graph relative to one of the edges (considered to be the *fixed edge*) is given by the mobility criterion

$$M = 2g - n_2 - 3$$

where g is the total number of vertices of the graph (regardless of their degree) and n_2 is the number of edges of the graph.

As with planar kinematic systems, the value obtained for M from the mobility criterion in Theorem 4.3 is always an integer. Again, M can be positive, zero or negative, and, by analogy with the planar kinematic systems, the *graph* is then said to be **mobile, immobile** or **overconstrained** respectively.

Here the significance of negative mobility is simply that the graph is the direct graph of an overconstrained kinematic system. Thus a graph with mobility –6 could lose up to six constraints (that is, edges) and still remain immobile. Correspondingly, the planar kinematic system it represents could lose up to six binary links without becoming mobile.

Table 4.1 illustrates some common graphs and their mobilities, derived using the criterion from Theorem 4.3.

Table 4.1 Mobilities of some common graphs treated as the direct graphs of planar kinematic systems.

	graph	g (vertices)	n_2 (edges)	M (mobility)
K_3		3	3	0
C_4		4	4	1
K_4		4	6	–1
C_6		6	6	3
$K_{3,3}$		6	9	0
octahedron		6	12	–3
K_6		6	15	–6
cube		8	12	1

4.4 Planar systems with mobility 0

We can now enumerate some planar kinematic systems with a particular mobility. In this subsection, *we concentrate on systems in which the only links present are binary links, and for which the mobility is 0*. These form an important class of kinematic system since they are used frequently as mechanical structures. In this case we have $n_r = 0$ for all values of r other than $r = 2$, and, since $M = 0$, the mobility criterion from Theorem 4.2 becomes

$$2g = n_2 + 3$$

where g is the total number of joints and n_2 is the number of binary links. The left-hand side of this equation is always an even number, so the right-hand side must also be even, and so n_2 must be odd. We can therefore tabulate the possible pairs of values for g and n_2 as follows:

g	3	4	5	6	7	8	9	10	11	12	...
n_2	3	5	7	9	11	13	15	17	19	21	...

This sequence of pairs of values for g and n_2 gives rise to a sequence of planar systems with mobility 0. In general, for each pair of values for g and n_2, there are several possible distinct physical arrangements of the links and joints.

We represent such a system by a *direct* graph. We continue to assume that each r-ary joint is equivalent to $r - 1$ revolute pairs.

Notice that, as the number of joints increases by 1, the number of binary links increases by 2.

Table 4.2   Planar systems with mobility 0, involving only binary links, represented by their direct graphs.

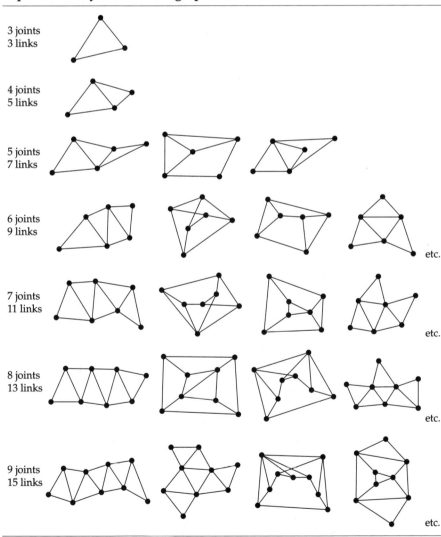

3 joints
3 links

4 joints
5 links

5 joints
7 links

6 joints
9 links
etc.

7 joints
11 links
etc.

8 joints
13 links
etc.

9 joints
15 links
etc.

Table 4.2 illustrates the direct graphs of some of the systems in the sequence. As you can see, the sequence includes not just triangulated systems

— some of them contain non-triangular subsystems. This can lead to certain difficulties. For example, the mobility of a system with 6 joints and 9 binary links ought to be 0, according to Theorem 4.2. However, it is possible to construct a system of 6 joints and 9 binary links whose mobility is 1. This happens if too many constraints are concentrated in one part of the system, leaving another part with too much freedom. The following diagram illustrates this situation.

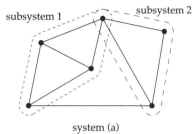
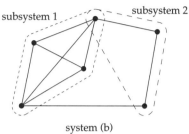

subsystem 1 · subsystem 2 · subsystem 1 · subsystem 2

system (a)   system (b)

Each of these systems has 6 joints and 9 links, and therefore should have mobility 0, according to Theorem 4.2. However, although system (a) is a rigid structure, system (b) is not, and has a single freedom. This happens because we have removed a link from subsystem 2 and added an extra diagonal link to subsystem 1. If we treat each subsystem in isolation, we see that subsystem 1 has *decreased* its mobility from $M = 0$ to $M = -1$, as predicted by Theorem 4.2. Correspondingly, subsystem 2 has *increased* its mobility from $M = 0$ to $M = 1$. We say that subsystem 1 is now *overconstrained* and subsystem 2 is *underconstrained*.

Initially subsystem 1 had 4 joints and 5 links, and now it has 4 joints and 6 links. Initially subsystem 2 had 3 joints and 3 links, and now it has 3 joints and 2 links.

Although these effects appear to cancel out when we apply the mobility criterion from Theorem 4.2 to the whole system, they do not cancel out in the real system because of the way the two subsystems are joined together: the underconstrained part dominates and gives the whole system mobility 1. This phenomenon generally always occurs if we overconstrain or underconstrain a part of our system, and it provides yet another example of the type of complication that can arise in a mobility calculation. It also provides a good argument for constructing a system in stages, so that at each stage the mobility is 0 (or whatever other value we wish to have).

4.5   Planar systems with mobility 1

So far we have enumerated some kinematic systems containing only binary links but having multiple joints. We now look at systems in which the links can be of any type (unary, binary, ternary, ... , r-ary), but the joints are all binary joints. *We concentrate on systems in which the only joints present are binary joints, all of which are revolute pairs, and for which the mobility is 1.* These form an important class of kinematic systems, since they are used frequently as mechanical linkages and mechanisms which require only one input to control them. For these systems it is simpler to use the mobility criterion from Theorem 4.1, rather than that from Theorem 4.2.

We represent such a system by an *interchange* graph.

If the mobility is 1, the mobility criterion from Theorem 4.1 becomes

$$3n = 2j + 4 \qquad (4.7)$$

where n is the total number of links and j is the number of revolute pairs.

The right-hand side of this equation is always an even number, and so n must be even. We can therefore tabulate the possible pairs of values for n and j as follows:

n	2	4	6	8	10	12	14	16	18	...
j	1	4	7	10	13	16	19	22	25	...

Notice that, as the number of links increases by 2, the number of binary joints increases by 3.

We shall now use the information in this table as a springboard for investigating the various *types* and *arrangements* of links that can form a system of n links and j binary joints (revolute pairs) with mobility 1. We investigate the systems corresponding to the first few entries in the table.

Systems with 2 links and 1 joint

It is easy to see that there is only one system containing two links and one revolute pair. This is just the simple hinge — a revolute pair connecting two unary links.

Systems with 4 links and 4 joints

Problem 4.4

Determine the number of possible arrangements of a planar kinematic system with four links and four revolute pairs.

Hint Determine the possible *interchange* graphs of such a system.

We may obtain all possible four-link arrangements by using equation 4.2. For a system with four links, this becomes

$$n_0 + n_1 + \cdots + n_r + \cdots + n_s = 4 \tag{4.8}$$

For this equation to have positive integer solutions, each term must be less than or equal to 4, and since 4 cannot be partitioned into more than four positive integers $(1 + 1 + 1 + 1)$, there are at most four non-zero terms on the left-hand side.

To determine which of the terms on the left-hand side can be non-zero, we first consider what we can deduce from the fact that the only joints are revolute pairs. For a four-link system there are at most four such joints. But there can be no quaternary link, since if the four joints were all on one link, there would need to be at least *five* links present in all (say, four unary links attached to a central quaternary link). Similarly there can be no r-ary link for any $r > 4$, and consequently the maximum multiplicity of any link in our four-link system is 3. Thus we have $n_4 = n_5 = \ldots = n_s = 0$, and equation 4.8 becomes

$$n_0 + n_1 + n_2 + n_3 = 4 \tag{4.9}$$

We may also exclude nullary links because they imply a disconnected system, and unary links because they absorb the only available single freedom of a system with mobility 1, giving us nothing essentially different from a simple hinge. So we have $n_0 = n_1 = 0$, and equation 4.9 reduces further to

$$n_2 + n_3 = 4 \tag{4.10}$$

Finally, from equation 4.6 we have

$$2n_2 + 3n_3 = 2j \tag{4.11}$$

since $j = g$ if all joints are binary.

The only simultaneous solution of equations 4.7, 4.10 and 4.11 for $n = 4$ is $j = 4$, $n_2 = 4$ and $n_3 = 0$. The only planar four-link system with mobility 1 and involving only revolute pairs is thus the system of four binary links connected by four revolute pairs given in the solution to Problem 4.4.

Systems with 6 links and 7 joints

The next pair of entries in the table is $n = 6$, $j = 7$. How many distinct systems are there with six links, seven joints and mobility 1?

We begin by following the same procedure as for the four-link system. Thus we write equations 4.2 and 4.6 in the appropriate forms, taking account of the total number of links and binary joints, so that $n = 6$ and $j = g = 7$. This gives the two equations

$$n_2 + n_3 + n_4 + \cdots + n_r + \cdots + n_s = 6 \qquad (4.12)$$

$$2n_2 + 3n_3 + 4n_4 + \cdots + rn_r + \cdots + sn_s = 14 \qquad (4.13)$$

where we have eliminated the nullary and unary links for the same reasons as before.

Positive integer solutions to equation 4.12 imply that there can be at most six non-zero terms. We can further restrict the terms by noting that there cannot be more joints on any one link than the total number of joints in the system. This places an upper bound of 7 on the multiplicity s of the links, so that $n_r = 0$ for $r > 7$. However, we can do better than this, since in any system of only six links, no link can be connected to more than five others, so that 5 is the highest possible multiplicity for a link in this system, and hence $n_r = 0$ for $r > 5$.

Furthermore, since, as in the case of systems with 4 links and 4 joints, we do not allow unary links, we can eliminate systems of type (a) below, and so $s = 4$ is the maximum multiplicity for a link.

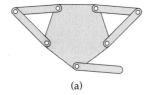 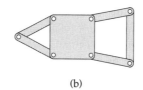

(a)   (b)

We can also stipulate that there be no *separating* link in the system. This is a link whose removal would disconnect the system. For example, system (b) above has a quaternary separating link, and we discount such systems because the various subsystems meeting at the separating link are essentially kinematically independent from one another. For a system with mobility 1, this means that one subsystem can exhibit this mobility, while all other subsystems have mobility 0. Thus, in system (b), there is a triangular circuit forming a subsystem to the left of the quaternary link. This circuit is a subsystem with mobility 0, so that the two binary links involved are immobile. Essentially, the triangular circuit can be absorbed into the quaternary link, which then becomes a binary link, so that the whole system behaves just as a four-link system with mobility 1.

If we translate the above observations into graph terminology, we see that the interchange graph of a system without separating links is a connected graph which remains connected after the removal of any vertex. For a graph of this type, with n vertices and j edges, the degree r of any vertex must satisfy the inequality

$$r \le j - n + 2$$

You were asked to prove this result in Problem 1.9.

Furthermore, if the graph represents a system with mobility 1, then we can substitute for j from equation 4.7 and obtain the simpler inequality

$$r \le n/2 \qquad (4.14)$$

If we now return to our system with 6 links, 7 joints and mobility 1, and stipulate that the interchange graph should be connected, and should remain connected after the removal of any vertex, we have, from equation 4.14, that $r \le 3$, and so ternary links are the most complex type that can occur. Equations 4.12 and 4.13 thus become:

$$n_2 + n_3 = 6$$

$$2n_2 + 3n_3 = 14$$

Solving these two equations simultaneously for n_2 and n_3, we obtain the values $n_2 = 4$, $n_3 = 2$. Our system therefore must consist of four binary links and two ternary links. These can be assembled in two ways, so that we get two distinct systems. They are both shown below, together with their interchange graphs.

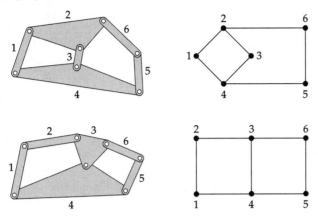

Systems with 8 links and 10 joints

Next, we enumerate the possible systems with eight links, ten joints and mobility 1, whose interchange graphs are connected, remain connected after the removal of any vertex (so that there are no separating links) and have only vertices of degree greater than or equal to 2 (so that there are no nullary or unary links).

We begin by noting that $n_0 = n_1 = 0$ (since there are no nullary or unary links) and that $r \leq 4$ (by inequality 4.14), so that such a system can have only binary, ternary and quaternary links in it. Equations 4.2 and 4.6 become:

$$n_2 + n_3 + n_4 = 8$$

$$2n_2 + 3n_3 + 4n_4 = 20$$

There are only three different partitions of 8 that satisfy both of these equations, and these are tabulated below.

n_2	n_3	n_4	
4	4	0	4 binary links, 4 ternary links
5	2	1	5 binary links, 2 ternary links, 1 quaternary link
6	0	2	6 binary links, 2 quaternary links

These three partitions give rise to sixteen distinct systems, because the first can be assembled in nine different ways, the second in five different ways, and the third in two different ways. We illustrate the systems, together with their interchange graphs, in Table 4.3.

Systems with more than 8 links

We have seen that, for planar systems with mobility 1 and involving only binary joints, all of which are revolute pairs: there is one such system with 2 links and 1 joint, there is one such system with 4 links and 4 joints, there are two such systems with 6 links and 7 joints, and there are sixteen such systems with 8 links and 10 joints. When we consider systems with 10 links and 13 joints, the combinatorial explosion starts to take over, and we discover 230 such systems. Beyond this, it has been shown that there are 6856 such systems with 12 links and 16 joints.

Table 4.3 The sixteen eight-link planar kinematic systems with mobility 1 involving only revolute pairs.

system	interchange graph

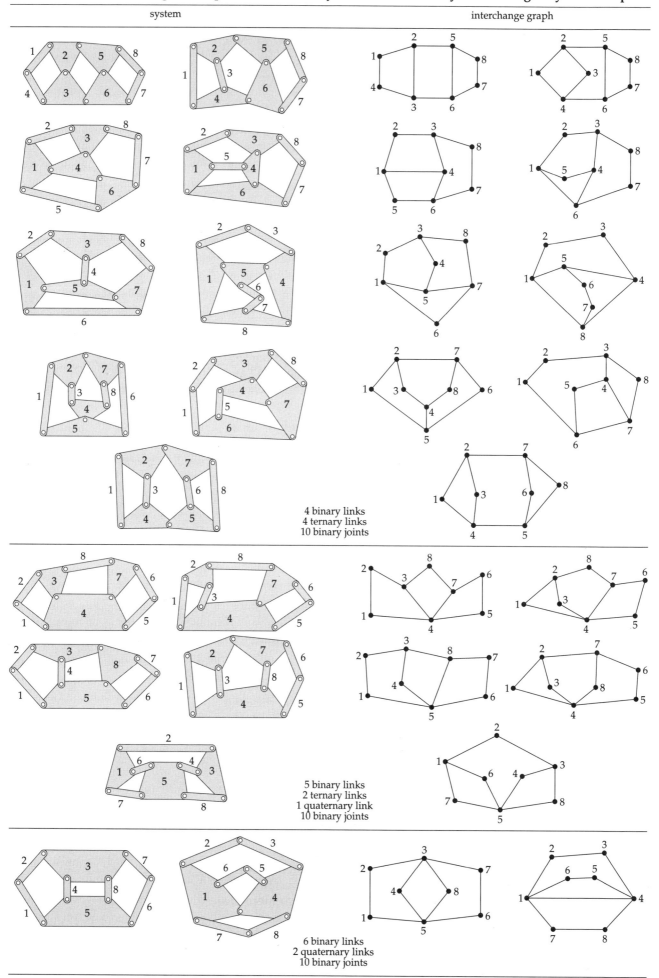

4 binary links
4 ternary links
10 binary joints

5 binary links
2 ternary links
1 quaternary link
10 binary joints

6 binary links
2 quaternary links
10 binary joints

4.6 Other planar and spatial systems

We have now discussed two types of planar kinematic system involving only revolute pairs — those with mobility 0 and those with mobility 1. Planar kinematic systems involving other Reuleaux pairs and/or with mobilities greater than 1 can be analysed using a similar approach without much extra difficulty.

More general types of planar kinematic system have prismatic and/or planar pairs as well as revolute pairs. For instance, the slider-crank mechanism is an example of a planar kinematic system having both revolute and prismatic pairs. The mobility of this type of system is calculated in the same way as before, with each link contributing three freedoms initially, and each joint (whether revolute or prismatic) imposing two constraints. This is possible because revolute and prismatic pairs both have connectivity 1. Theorem 4.1 may then be used with j representing the total number of revolute and prismatic pairs combined. The mobility of the slider-crank mechanism is therefore given by

$$
\begin{aligned}
M &= 3(4-1) - (2 \times 4) \\
  &= 9 - 8 \\
  &= 1
\end{aligned}
$$

Similar remarks apply to the polisher system. This is an example of a planar kinematic system having both revolute and planar pairs. Again the mobility of this type of system is calculated by counting initially three freedoms for each link, and then subtracting the appropriate number of constraints imposed by each joint. Each revolute pair imposes two constraints as before, but each planar pair imposes no constraints in two-dimensional space because its connectivity is 3. So the number of constraints imposed by the revolute pairs must be counted separately from the number of constraints imposed by the planar pairs in Theorem 4.1. The mobility of the polisher system is therefore given by

$$
\begin{aligned}
M &= 3(4-1) - (2 \times 3) - (1 \times 0) \\
  &= 9 - 6 - 0 \\
  &= 3
\end{aligned}
$$

If we were to generalize even further and investigate *spatial* kinematic systems, we would face several additional problems. For instance, the mobility criteria in Theorems 4.1 and 4.2 would be more complicated since they would have to take account of the six freedoms of a link in space, and would involve separate terms for the constraints imposed by each of the six Reuleaux pairs that might be present. Moreover, the range of possible geometric complications would increase significantly since there would be more scope for joint axes to be parallel, perpendicular or intersecting in various combinations.

In Section 5, we introduce two important types of spatial kinematic system, each with mobility 6, which are used in practical situations. These are a robot manipulator arm and a flight simulator platform.

4.7 Braced rectangular frameworks revisited

In Section 2 we saw how braced rectangular frameworks can be represented by bipartite graphs, and we saw that a framework is rigid if and only if the corresponding bipartite graph is connected. Moreover, we saw that a framework is minimally braced if and only if its bipartite graph is a spanning tree. But how does this relate to our discussion of mobility? In particular, what is the mobility of a braced rectangular framework? This question can be answered if, instead of using a rectangular plate to brace a bay in a framework, we use a link across adjacent sides, opposite sides or opposite corners of the bay.

Before being braced, an isolated bay consists of four binary links interconnected in a closed loop by four binary joints (revolute pairs). The mobility of the bay can be calculated from the mobility criterion in Theorem 4.1, with $n = 4$ and $j = 4$. This gives

$$M = 3(n - 1) - 2j$$
$$= 3(4 - 1) - (2 \times 4)$$
$$= 1$$

Since the bay has mobility 1 it is mobile and hence not rigid, just as we would expect from our earlier discussion.

If a brace is now inserted across two adjacent links in the bay, we introduce one extra binary link (the brace), and two extra binary joints (revolute pairs), located where this extra link joins the two adjacent links. We also convert these two adjacent binary links into ternary links.

3 binary links
2 ternary links
6 binary joints

3 binary links
2 ternary links
6 binary joints

5 binary links
2 binary joints
2 ternary joints

We now have three binary links, two ternary links and six binary joints, so $n = 5$ and $j = 6$. The mobility criterion from Theorem 4.1 gives

$$M = 3(n - 1) - 2j$$
$$= 3(5 - 1) - (2 \times 6)$$
$$= 0$$

So the isolated braced bay has zero mobility and hence is rigid, as expected.

Instead of using Theorem 4.1 we could have used Theorem 4.2, with $n_2 = 3$, $n_3 = 2$ and $g = 6$. This gives

$$M = 2g - n_2 - 3n_3 - 3$$
$$= (2 \times 6) - 3 - (3 \times 2) - 3$$
$$= 0$$

Again, we find that the isolated braced bay has zero mobility.

A similar situation arises if we insert the brace across two opposite links in the bay. This introduces one new binary link, and two extra binary joints (revolute pairs), located where this extra link joins the two opposite links. It converts these two opposite binary links into ternary links, as before, and we again have $n = 5$ and $j = 6$, giving a mobility of zero (by Theorem 4.1).

Alternatively, $n_2 = 3$, $n_3 = 2$ and $g = 6$ give a mobility of zero (by Theorem 4.2).

A complication can arise if the extra binary link forming a brace between a pair of opposite links in a bay is parallel to the other pair of opposite links. In this case no bracing is achieved since the geometry results in a five-link system with full-cycle mobility, and so the mobility is not reduced.

If we insert the diagonal brace joining opposite corners of the bay, the brace changes two of the binary joints into ternary joints, without converting any binary links into ternary links. We are now faced with a choice when we wish to calculate the mobility.

First, we can treat all of the links as binary links and count the number of joints regardless of their multiplicity. This gives $n_2 = 5$ and $g = 4$, and so, by Theorem 4.2, the mobility is

$$M = 2g - n_2 - 3$$
$$= (2 \times 4) - 5 - 3$$
$$= 0$$

This again confirms the rigidity of the bay.

Alternatively, we can expand each ternary joint into two binary joints. This allows us to use the simpler mobility criterion from Theorem 4.1. We show one way of performing this expansion in the following diagram.

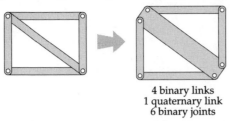

4 binary links
1 quaternary link
6 binary joints

Here we have expanded the two ternary joints in such a way that the diagonal brace (which was previously a binary link) has become a quaternary link, while all joints are now binary, and four of the original binary links remain. So we have $n = 5, j = 6$ and the mobility criterion of Theorem 4.1 gives

$$M = 3(n-1) - 2j$$
$$= 3(5-1) - (2 \times 6)$$
$$= 0$$

So we have zero mobility, and hence a rigid bay, as we would expect.

In general, a ternary joint can be expanded in three different ways, so, for an isolated bay braced with a single diagonal brace, there are *nine* different ways to expand the *two* ternary joints.

See Problem 1.4.

Problem 4.5

Sketch four of the remaining eight different ways of expanding the two ternary joints of an isolated rectangular bay braced with a simple diagonal brace joining opposite corners. In each case, state the number of binary joints and the number and type of links so produced, and hence calculate the mobility using Theorem 4.2.

As yet, we have only dealt with a single isolated braced bay. But most frameworks consist of more than one bay. In an unbraced rectangular framework with two bays there are seven binary links, four binary joints and two ternary joints. So even before braces are introduced into the framework, we must deal with ternary joints. If two diagonal braces joining opposite corners are inserted then we already have the possibility of a 5-ary joint, since the diagonal braces may be in different directions. And in a larger framework with such diagonal braces it is possible to have 8-ary joints (the joints not on the boundary of the framework are 4-ary even before braces are inserted).

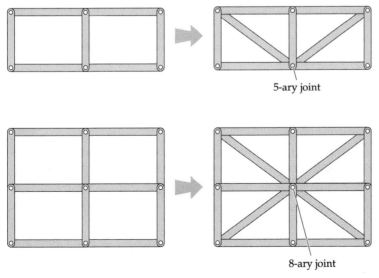

5-ary joint

8-ary joint

If the bays are diagonally braced, the number of ways to expand the multiple joints grows rapidly with the number of braced bays in the framework. The following diagrams show a 2×2 braced rectangular framework together with one possible expansion of its multiple joints.

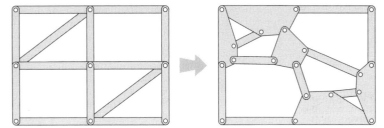

In determining the mobility of braced rectangular frameworks, it is often easier to use the mobility criterion from Theorem 4.2 directly rather than to expand the multiple joints first. Furthermore, since only binary links occur together with multiple joints, this criterion reduces to the simpler form in Theorem 4.3, namely

$$M = 2g - n_2 - 3,$$

with n_2 being the total number of binary links and g being the total number of joints regardless of multiplicity.

After studying this section, you should be able to:

- derive a simple mobility criterion for planar systems, based on combinatorial considerations, and appreciate the types of geometrical complication that can arise;

- use appropriate variations of the mobility criterion in terms of multiple links or multiple joints, as required;

- sketch the direct graphs of some planar systems with mobility 0;

- sketch some examples of planar systems with up to eight links and mobility 1;

- reproduce a system from its interchange graph;

- determine the mobility of simple braced rectangular frameworks.

5 Flight simulators and robots

This section comprises the television programme related to this unit. The programme is described in the related *Television Notes*.

Further reading

There is no single book that deals with kinematics from a purely combinatorial viewpoint. However, several books deal with parts of the subject in an interesting way.

An excellent book on kinematic geometry is:

K. H. Hunt, *Kinematic Geometry of Mechanisms*, Clarendon Press, 1978.

This is an advanced text, although it is well written and explains many difficult ideas with good clear diagrams. The first chapter on the components of mechanism and the second chapter on freedom and structure in mechanism are particularly relevant.

For a more traditional approach to kinematic systems, and especially planar kinematic systems, read:

R. S. Hartenberg and J. Denavit, *Kinematic Synthesis of Linkages*, McGraw-Hill, 1964.

This contains an interesting outline of the history of kinematics up to 1900, as well as a useful chapter on making kinematic models.

The original classic text presenting the first systematic approach to the description of machinery in kinematic terms is:

F. Reuleaux, *The Kinematics of Machinery*, Dover Publications, 1963 (reprint of the original Macmillan edition of 1876).

This book is still relevant to modern kinematic theory, and indeed it introduced the important concepts of higher and lower kinematic pairs, as well as many other lasting ideas. It is well worth reading.

A useful source book for ideas on the design of kinematic systems is:

N. P. Chironis, *Mechanisms, Linkages and Mechanical Controls*, McGraw-Hill, 1965.

The presentation is mainly pictorial, consisting of many reprinted articles describing actual systems in use.

Finally, for an account of industrial robot technology, read:

J. F. Engelberger, *Robotics in Practice: Management and Applications of Industrial Robots*, Kogan Page, 1980.

Acknowledgement

p.18 picture of Reuleaux, courtesy of Deutsches Museum, Munich.

Exercises

Section 1

1.1 The diagram in the margin illustrates the links and joints on one end of a cantilevered toolbox or sewing basket, with the carrying handles omitted for clarity. Assuming that the other end has the same number and arrangement of links and joints, determine, giving their multiplicity:

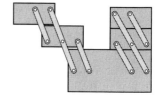

(a) all the links in the system;

(b) all the joints in the system.

1.2 Draw the interchange graph of one end of the cantilevered box shown in Exercise 1.1, and indicate why a direct graph representation is inappropriate for this system.

1.3 The diagram in the margin illustrates a typical telescopic radio aerial, which is freely pivoted so that it can point in any direction above the plane on which it is mounted.

(a) How many links and joints does the system have?

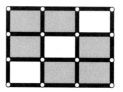

(b) What types of joint are present?

1.4 A *surface of revolution* is obtained by revolving a curve about an axis in the plane of the curve. A cone is produced in this way from a straight line, and this surface can be used as the surface of contact for the two links in a revolute pair.

(a) Give one other example of a curve and its surface of revolution which is suitable for use in a revolute pair.

(b) Give one example of a curve and its surface of revolution which is *not* suitable for use in a revolute pair, and explain why not.

Section 2

2.1 The diagram in the margin illustrates a 3×3 rigid braced rectangular framework.

(a) Explain why the framework is rigid.

(b) List those bays whose brace may be removed without altering the rigidity.

Section 3

3.1 Suppose that a suitably shaped rigid body remains in complete surface contact with both the plane and the indentation shown in the following diagrams. How many freedoms does the body have in each case?

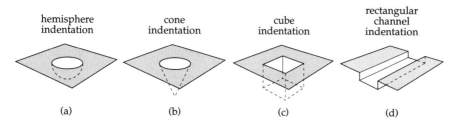

hemisphere indentation	cone indentation	cube indentation	rectangular channel indentation
(a)	(b)	(c)	(d)

3.2 How many points of a rigid body can be constrained by a general surface so that, if any further points are so constrained, then no more freedoms are removed?

3.3 A thin triangular slab lies on a plane surface. Two of its corners are constrained to lie on different, non-parallel straight lines in the plane. State how many freedoms the constrained slab has, and indicate suitable quantities or coordinates to describe its (two-dimensional) pose.

Section 4

4.1 Determine the mobility of the cantilevered box in Exercise 1.1. (Treat the system as a planar kinematic system and ignore those links and joints hidden from view on the other end of the box.)

4.2 Describe the kinematic structure of a deckchair frame when it is (a) erected and (b) not erected, and state its mobility as a planar kinematic system in each case.

4.3 The diagram in the margin illustrates a cross-section of a simple design for the supporting frame of an umbrella. By treating this cross-section as a planar kinematic system:

(a) state the number of links present and indicate their multiplicities;

(b) state the number of joints present, indicate their multiplicities, and state which Reuleaux pairs they are;

(c) determine the mobility of the system.

Hint Remember that a prismatic pair constrains *two* freedoms in the plane, and so the theorems of Section 4.3 can be extended to include such pairs as well as revolute pairs.

4.4 The diagram in the margin illustrates the vertical cross-section of the structural frame of a mansard roof. The cross-section can be considered as a planar kinematic system consisting of four links connected at three revolute pairs, B, C and D. The points A and E indicate where the roof rests on the walls of the building. We want to make the roof rigid so that it 'sits' on the tops of the walls without exerting any sideways forces.

Mansard roofs are named after the French architect François Mansard.

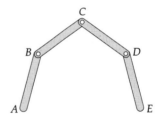

(a) What is the mobility of the above four-link, three-joint system?

(b) If extra binary links are to be attached by revolute pairs at A, B, C, D or E, how many are required to make the roof rigid?

(c) Sketch at least three of the possible rigid systems that can be formed in this way.

Section 5

5.1 Give three examples of objects or devices which utilize a kinematic system to position and orientate them in three-dimensional space. Indicate the amount of freedom required by each of your examples in order for it to function adequately.

Solutions to the exercises

1.1

(a) There are five trays (two are quaternary links, two are 6-ary links, and one is an 8-ary link) and twelve connectors (eight are binary links and four are ternary links).

(b) There are 14 binary joints at each end — that is, a total of 28 binary joints.

1.2 The interchange graph of just one end of the box is shown in the margin, where t denotes a tray. (It is drawn so that the vertices and edges of the graph correspond approximately to the spatial positions of the links and joints in the mechanical system when the box is open.)

A direct graph is inappropriate because there are links present with multiplicity greater than 2.

1.3

(a) 5 links, 4 joints.

(b) 4 binary joints comprising 1 spherical pair and 3 cylindric pairs.

1.4

(a) An ellipse is a suitable curve. If it is revolved about either its major axis or its minor axis, it produces an ellipsoid and, as a contact surface, this allows only rotation about the chosen axis.

(b) A circle revolved about one of its diameters is not suitable for a revolute pair, because the resulting surface is a sphere, and this allows rotations about more than one axis.

2.1

(a) The framework is rigid because the associated bipartite graph is connected.

(b) The bipartite graph is a 6-cycle, and *any one* of its six edges can be removed without disconnecting the graph. So, *any one* brace can be removed without altering the rigidity. The six bays from which a brace can be removed are therefore bay(1, 1), bay(1, 2), bay(2, 1), bay(2, 3), bay(3, 2) or bay(3, 3).

3.1

(a) The hemisphere by itself allows rotations about three mutually perpendicular axes. However, the plane prevents two of these. Therefore there is only one possible rotation (about an axis perpendicular to the plane) and thus the body has just one freedom.

(b) The surface of the cone by itself allows only rotation about the axis of the cone. If this axis is perpendicular to the plane, then the body has one freedom. However, if the axis of the cone is not perpendicular to the plane, then the body has zero freedoms.

(c) The surface consisting of five faces of the cube prevents all motion. Thus the body has zero freedoms.

(d) The rectangular channel by itself allows a translation. If the plane is parallel to the bottom of the channel or intersects it in a line parallel to the axis of the channel, then the body has one freedom. Otherwise it has zero freedoms.

3.2 If the surface is a plane, then three points can be constrained (a three-legged stool). If the surface has 'bumps', 'ridges', 'dimples' or 'channels', then generally more points can be constrained. For example, a cylinder allows four points to be constrained. Since a body in space has six freedoms initially, then the maximum number of points that can be constrained by any surface is six, because each constrained point removes one freedom.

3.3 A slab free to move on a plane surface has three freedoms (see Problem 3.5(a)). Constraining two of its corners to lie on different, non-parallel straight lines in the plane removes two of these freedoms, leaving the slab with only one freedom.

A suitable quantity to describe its pose is the distance d of one vertex A from a fixed point along the straight line on which A is constrained to lie.

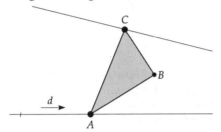

4.1 Each joint is a revolute pair. Using Theorem 4.1 with $n = 11$ and $j = 14$, we have

$$M = 3(n - 1) - 2j$$
$$= (3 \times 10) - (2 \times 14)$$
$$= 2$$

4.2

(a) If the deckchair is erected, then its kinematic structure consists of a triangular framework arrangement composed of three binary links and three binary joints, all of which are revolute pairs. If the system is considered as a planar kinematic system, then by Theorem 4.1, with $n = 3$ and $j = 3$, we have

$$M = 3(n - 1) - 2j$$
$$= 3(3 - 1) - (2 \times 3)$$
$$= 0$$

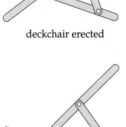

deckchair erected

(b) If the deckchair is not erected, then its kinematic structure consists of two unary links, one binary link and two binary joints (both of which are revolute pairs). So $n_0 = 0$, $n_1 = 2$, $n_2 = 1$, $n_r = 0$ for $r \geq 3$ and $g = 2$. If the system is considered as a planar kinematic system, then by Theorem 4.2 we have:

$$M = 2g + n_1 - n_2 - 3$$
$$= (2 \times 2) + 2 - 1 - 3$$
$$= 2$$

deckchair not erected

4.3

(a) There are four binary links (the 'folding' parts) and two ternary links (the stem and the slider) present in the system, so $n_0 = n_1 = 0$, $n_2 = 4$, $n_3 = 2$ and $n_r = 0$ for $r \geq 4$.

(b) There are seven binary joints present, so $g = 7$. Six joints are revolute pairs and one joint is a prismatic pair.

(c) In a planar kinematic system each link has three freedoms when unconstrained, each revolute pair constrains two freedoms, and each prismatic pair constrains two freedoms. So, by extending Theorem 4.2, we have

$$M = 2g - n_2 - 3n_3 - 3$$
$$= (2 \times 7) - 4 - (3 \times 2) - 3$$
$$= 1$$

It is not difficult to find general surfaces such that a six-legged stool (with its feet in contact) cannot move in any way without lifting one leg clear of the surface. For example, the corner of a box provides a suitable surface with two legs of the stool resting on each of the three planes meeting at the corner.

Theorem 4.2 can be extended to cover all joints that constrain two freedoms, and so can be used for prismatic as well as revolute pairs (though not for planar pairs).

4.4

(a) Using Theorem 4.1 with $n = 4$ and $j = 3$, we obtain

$$M = 3(n - 1) - 2j$$
$$= 3(4 - 1) - (2 \times 3)$$
$$= 3$$

(b) We first note that we need to constrain the point A with a binary link joining it to C, D or E. Similarly, we need to constrain the point E with a binary link joining it to A, B or C. Hence AB and DE must become binary links.

We can now use Theorem 4.2 to help us determine the number of binary links needed to make the mansard frame rigid. For the original mansard frame we have $g = 3$, $n_1 = 2$ and $n_2 = 2$, so that

$$M = 2g + n_1 - n_2 - 3$$
$$= (2 \times 3) + 2 - 2 - 3$$
$$= 3$$

If we now turn AB and DE from unary links into binary links, without for the moment considering how many binary links we are adding, we create two new joints, at A and E, so that g increases from 3 to 5; we also decrease the number n_1 of unary links from 2 to 0 and increase the number n_2 of binary links from 2 to 4, so that

$$M = 2g - n_2 - 3$$
$$= (2 \times 5) - 4 - 3$$
$$= 3$$

and there is no overall effect on the value of M.

Since we cannot increase g beyond 5, because the only possible joint positions are at A, B, C, D and E, we must further increase n_2 by 3 in order to decrease the mobility to 0 (and therefore make the mansard frame rigid). In other words, we must add three binary links to make the mansard frame rigid.

(c) There are several possibilities — for example:

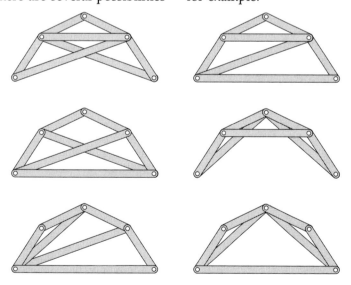

5.1 There are many possible examples. Three representative types are:

(a) a television camera;

(b) a spotlight;

(c) a paint spray-gun.

In each case, the requirement is that of positioning the device at any location within a certain volume of space (a workspace), and orientating the lens axis, the light beam or the spray-nozzle to point in a certain angular range of directions from that location. Positioning the device necessitates the fixing of three coordinates, such as its distances from two perpendicular walls, together with its height above the floor. Orientating the device necessitates the fixing of two more coordinates, such as the pan and tilt angles.

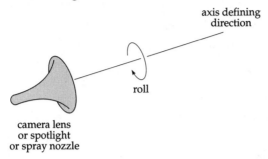

axis defining
direction

roll

camera lens
or spotlight
or spray nozzle

Thus we require five coordinates to specify the arrangement and we say that these systems require five freedoms.

In general, spatial kinematic devices would require six freedoms, because an additional flexibility is needed. The systems that we have chosen need only five freedoms because they have an axis associated with their function, and rotation about this axis (that is, a freedom in roll angle) is superfluous to their main function. An exception might be made for the television camera, which is sometimes required to perform a roll.

Solutions to the problems

Solution 1.1

Possibilities include:

(a) opening window, door, door handle, door lock, lavatory seat, tap, shower head, shower controls, central-heating pump, loft ladder, electric light switch;

(b) retracting undercarriage, aileron, flap, rudder, elevator, joystick, swing-wing, propeller, engine, rotor;

(c) boom, rudder, lifting or swing keel, self-steering gear, winch, radar antenna, compass;

(d) wheel, lifting foot-rest, brake lever, folding wheelchair frame;

(e) keyboard keys, disk drive, mouse, on–off switch, monitor brightness control knob.

Solution 1.2

Possibilities include:

(a) scissors — the blades are links, the pivot is a joint;
cupboard — the doors and the cupboard carcase are links, the hinges are joints;
cooker control panel — the knobs are links, their spindles are joints;
hi-fi turntable — the tone-arm is a link, its pivot is a joint;

(b) filing cabinet — the drawers and the cabinet body are links, the runners are joints;
swivel chair — the seat and its pedestal are links, the swivel is a joint;
dividers/compasses — the arms are links, the pivot is a joint;

(c) fork-lift truck — the wheels are links, their axles are joints, the fork is a link, its vertical slideway is a joint;
lathe — the cutting tool is a link, the spindle shaft and leadscrew are joints;

(d) roundabout — the platform is a link, the axis is a joint;
see-saw — the beam is a link, its pivot is a joint;

(e) lift — the lift doors are links, their slideways are joints;
revolving doors — the doors are links, their axis is a joint.

Solution 1.3

Possibilities include:

(a) a wheel, connected at a single joint, its axle;
a drawer, connected at a single joint, its runner;
a door handle, connected at a single joint, its spindle;

(b) the main body of a deckchair, connected at two separate joints to the two folding legs;
the body of a two-blade penknife, connected at two separate joints to the two folding blades;
a link in a bicycle chain, connected at two separate joints to the two adjacent links;

(c) a balance beam, connected at three separate joints to the frame and to each of its two pans;
a lever, connected at three separate joints to the load, the fulcrum, and the applied effort.

Solution 1.4

There are precisely three ways in which a ternary joint can be expanded into binary joints:

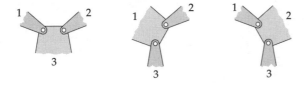

Solution 1.5

Possibilities include:

> the hinge of a suitcase;
> the pivot of the two blades of a pair of scissors;
> the hinge of a cosmetics compact case;
> the joint between the piston and tube of a bicycle pump;
> the joint between a bottle and its screw-top.

Solution 1.6

(a) 3; (b) 3; (c) 1; (d) 1; (e) 0; (f) 0.

Solution 1.7

If there is an r-ary link in a system, for $r \geq 3$, then it is not possible to represent it by an edge of a graph, since in a graph each edge joins just *two* vertices.

What we require is some way of representing the *multiplicity* of the link directly, and this can be done by allowing the 'edges' of a graph to be plane or space segments, in addition to the usual line segments. Thus a ternary link could be represented by a triangular plane segment, a quaternary link by a quadrilateral plane segment or tetrahedral space segment, and so on. However, the resulting direct 'graphs' are then not graphs at all, but are known as *hypergraphs*.

Solution 1.8

(a) Since the vertices represent joints and the edges represent links, a vertex of degree 3 in a direct graph represents a ternary joint, and in general a vertex of degree r represents an r-ary joint.

(b) Since the vertices represent links and the edges represent joints, a vertex of degree 3 in an interchange graph represents a ternary link, and in general a vertex of degree r represents an r-ary link.

Solution 1.9

Suppose we remove a vertex of degree r (representing an r-ary link) from an interchange graph with n vertices and j edges. This operation removes r edges; and $n - 1$ vertices remain. Since the removal of one vertex does not disconnect the graph, it remains at least a tree. A tree with $n - 1$ vertices has $n - 2$ edges, and so the original graph must have at least this number of edges, together with the r edges that were removed. We have therefore shown that $j \geq r + n - 2$, and rearranging this inequality we get $r \leq j - n + 2$.

Solution 1.10

The systems look like their direct graphs. All six are shown below.

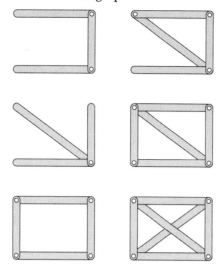

Solution 1.11

Revolute pair

Examples are provided by wheels turning on axles, door handles that turn, a pair of scissors or shears, the tuning knob on a radio receiver, a car steering-wheel and steering-column, clutch pedal and brake pedal, aircraft or ship propellers, and a turbine rotating on its shaft.

Prismatic pair

Examples are provided by sliding doors and their tracks, push-button switches on radios or televisions, keys on calculators or typewriters, camera shutter-release buttons, and nails inserted into timber, walls, etc.

Screw pair

Examples are provided by screw-tops on bottles and jars, water taps, leadscrews on lathes, car jacks, screw-in light bulbs, lipsticks, SLR (single lens reflex) camera lenses, and screws inserted into timber, walls, etc.

Cylindric pair

Examples are provided by bayonet-fitting light bulbs, agitators on some washing-machines (which rotate and slide up and down on their spindles), and syringes.

Planar pair

Examples are provided by a puck on ice, a sander and work surface, a polisher and floor, and the swash-plate joint found on the rotor-head mechanism of helicopters.

Planar pairs are not common as practical joints, and are often simulated by combinations of revolute and prismatic pairs.

Spherical pair

Examples are provided by the ball in a ball-point pen or roll-on deodorant, and the ball-and-socket type of artificial hip.

Solution 1.12

Sketches are given opposite for the arrangement of axes of revolute, prismatic and screw combinations.

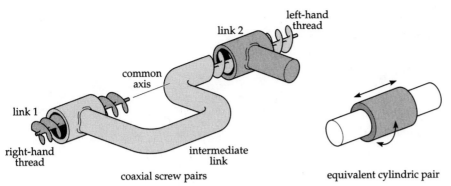

left-hand thread

link 2

common axis

link 1

right-hand thread

intermediate link

coaxial screw pairs

equivalent cylindric pair

(a) combination of two screw pairs equivalent to a cylindric pair

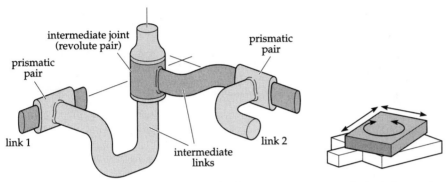

intermediate joint (revolute pair)

prismatic pair

prismatic pair

link 1

link 2

intermediate links

three screw pairs with mutually perpendicular axes

equivalent planar pair

(b) combination of two prismatic pairs and one revolute pair equivalent to a planar pair

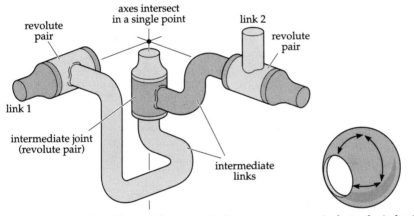

axes intersect in a single point

link 2

revolute pair

revolute pair

link 1

intermediate joint (revolute pair)

intermediate links

three screw pairs with mutually perpendicular axes

equivalent spherical pair

(c) combination of three revolute pairs equivalent to a spherical pair

Solution 1.13

An example of a spatial kinematic system found on a building site is provided by a mechanical excavator used for digging trenches, etc.

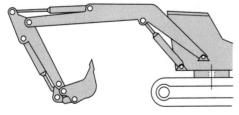

back-acting excavator

Solution 2.1

(a) Three binary links, two ternary links, and six binary joints.

(b) Three binary links, two ternary links, and six binary joints.

(c) Five binary links, two binary joints, and two ternary joints.

Solution 2.2

We see from the braces in the framework that:

r_1 must remain perpendicular to c_2 and c_3;
r_2 must remain perpendicular to c_1 and c_2;
r_3 must remain perpendicular to c_1.

We deduce that:

(a) r_1 must remain parallel to r_2 and r_2 must remain parallel to r_3, and so r_1 must remain parallel to r_3;

(b) c_2 must remain parallel to c_3 and c_1 must remain parallel to c_2, and so c_1 must remain parallel to c_3;

(c) since r_1 must remain perpendicular to c_2, and c_2 is parallel to c_1, r_1 must remain perpendicular to c_1;
similarly, r_2 must remain perpendicular to c_3, r_3 must remain perpendicular to c_2, and r_3 must remain perpendicular to c_3.

Since all three conditions in Theorem 2.1 are satisfied, the framework is rigid.

Solution 2.3

The three bipartite graphs are as follows:

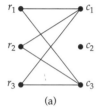 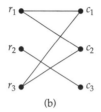 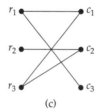

(a)   (b)   (c)

The bipartite graphs for frameworks (a) and (b) are disconnected, whereas the bipartite graph for framework (c) is connected.

Solution 2.4

r_1 parallel to r_3 $\leftrightarrow$ $r_1c_2r_2c_1r_3$
r_2 parallel to r_1 $\leftrightarrow$ $r_2c_2r_1$
r_2 parallel to r_3 $\leftrightarrow$ $r_2c_1r_3$
r_3 parallel to r_1 $\leftrightarrow$ $r_3c_1r_2c_2r_1$

c_1 parallel to c_2 $\leftrightarrow$ $c_1r_2c_2$
c_1 parallel to c_3 $\leftrightarrow$ $c_1r_2c_2r_1c_3$
c_2 parallel to c_1 $\leftrightarrow$ $c_2r_2c_1$
c_2 parallel to c_3 $\leftrightarrow$ $c_2r_1c_3$
c_3 parallel to c_1 $\leftrightarrow$ $c_3r_1c_2r_2c_1$
c_3 parallel to c_2 $\leftrightarrow$ $c_3r_1c_2$

r_1 perpendicular to c_1 $\leftrightarrow$ $r_1c_2r_2c_1$
r_1 perpendicular to c_3 $\leftrightarrow$ r_1c_3

r_2 perpendicular to c_2 $\leftrightarrow$ r_2c_2
r_2 perpendicular to c_3 $\leftrightarrow$ $r_2c_2r_1c_3$

r_3 perpendicular to c_2 $\leftrightarrow$ $r_3c_1r_2c_2$
r_3 perpendicular to c_3 $\leftrightarrow$ $r_3c_1r_2c_2r_1c_3$

Solution 2.5

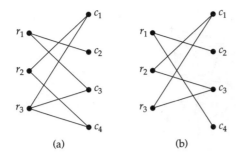

(a)　　　　　　　(b)

The bipartite graph for framework (a) is connected, so framework (a) is rigid. The bipartite graph for framework (b) is disconnected — the path $c_2 r_1 c_4$ and the cycle $c_1 r_3 c_3 r_2 c_1$ are components of the graph — so framework (b) is not rigid.

Solution 2.6

(a)

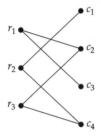

Since the bipartite graph is a spanning tree, the braced rectangular framework is minimally braced.

(b)　We find another spanning tree and construct the corresponding minimum bracing. For example:

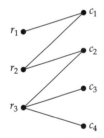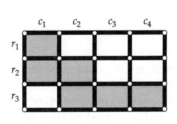

Solution 3.1

(a)　One possible alternative set of coordinates comprises the x- and y-coordinates of the vertex A, the y- and z- coordinates of the vertex B, and the z- and x- coordinates of the vertex C. We then have x_A, y_A, y_B, z_B, z_C, x_C as our six independent coordinates.

To see this, try specifying each of these in succession until the pose of the spacecraft is determined.

(b)　The six coordinates suggested are not all *independent*: specifying the values of any *five* of them fixes the position of the edge AB (as discussed in the text) and so automatically fixes the value of the sixth. But the spacecraft still has the freedom to rotate about the edge AB. A seventh coordinate would then be needed to specify the spacecraft's position completely.

Solution 3.2

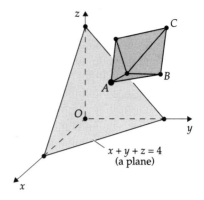

(a) vertex A lies on plane

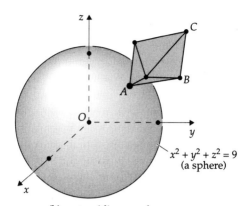

(b) vertex A lies on sphere

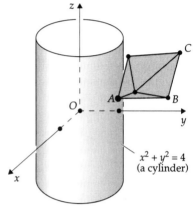

(c) vertex A lies on cylinder

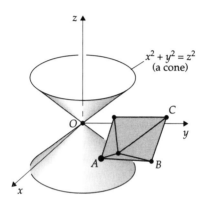

(d) vertex A lies on cone

Constraint equations:

(a) $x_A + y_A + z_A = 4$

(b) $x_A^2 + y_A^2 + z_A^2 = 9$

(c) $x_A^2 + y_A^2 = 4$

(d) $x_A^2 + y_A^2 = z_A^2$

Solution 3.3

The mobility is 4. This is because we can now describe the position of the spacecraft with just four coordinates. One of these could be the distance that A has moved along the orbit from some fixed point. The other three could be the same three angles θ_B, ϕ_B and ψ_C that we used earlier to describe the spacecraft's orientation.

Solution 3.4

We cannot make use of all eleven of the partitions, since we need to have at least three terms in the partition, so that we can constrain at least three points of the body (see Theorem 3.1). This eliminates partitions (a) (six constraints on one point), (b), (c) and (e) (constraints on only two points). We must also discount partition (d), because this places four constraints on one point, and we have already seen that a point can be immobilized with just three constraints. We are therefore left with partitions (f), (g), (h), (i), (j) and (k). These are all admissible.

Solution 3.5

(a) The slab has three freedoms. A typical set of three coordinates required to describe its pose comprises the x- and y- coordinates of one vertex A, and the angle that an edge AB makes with the x-direction. In this case, the coordinates are x_A, y_A, θ_B.

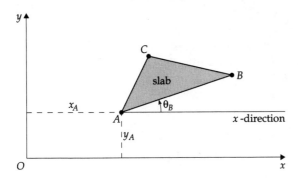

(b) Three constraints are required to immobilize the slab, since it has three freedoms. In two dimensions, a typical constraint is imposed by requiring a vertex, A say, of the slab to lie always on a particular *curve* (for example, a circle or straight line).

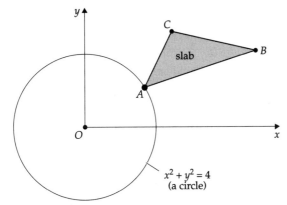

(c) Since at least two points (vertices A and B, say) are needed to describe the position of the slab, at least two points must be fixed in order to immobilize it. The vertex A can be fixed by two constraints, and a third constraint must be imposed on the second vertex, B, which will then also be fixed.

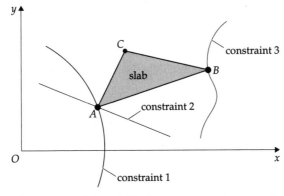

(d) It follows from parts (b) and (c) that we must impose *three* constraints on at least *two* points in order to immobilize the slab. The three possible partitions of 3 are:

3

2 + 1

1 + 1 + 1

We must eliminate the first partition because three constraints on one point are too many for a planar system of the type considered. This leaves the second and the third partitions as the only two possibilities. The second partition is the one we used above in part (c). The third partition is interesting in that it can be imposed by having one constraint on each of the three vertices of the triangular slab.

This is why we chose a triangle.

Solution 3.6

(a) If the contact surface is a cylinder, then we can constrain up to *four* non-collinear points, and therefore remove four freedoms (leaving a mobility of 2). For example, the rotational freedom about a line perpendicular to the surface is now lost, because it would raise at least one of the four points off the cylindrical surface. If we constrain a fifth point to lie on the cylinder (by adding another leg), then it does not remove the two translational freedoms.

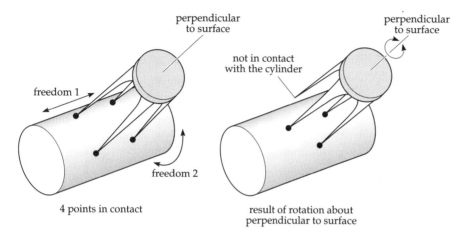

4 points in contact   result of rotation about perpendicular to surface

(b) If the contact surface is a sphere, then it is possible to constrain up to *three* distinct non-collinear points, and therefore to remove three freedoms (leaving a mobility of 3). The situation is similar to that for a plane surface, and the remaining three freedoms are illustrated below. There are two 'translational' freedoms (say, along a meridian of longitude and a parallel of latitude), together with one rotational freedom about a line perpendicular to the surface of the sphere. A fourth point constrained to lie on the sphere does not constrain any further freedoms.

This third freedom is sometimes referred to as *freedom in azimuth*.

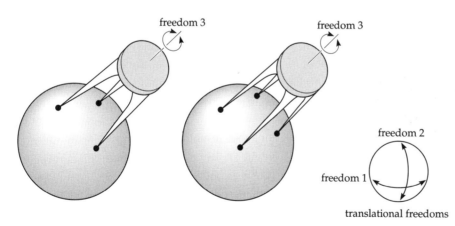

translational freedoms

(c) If the contact surface is a cone, then it is possible to constrain up to *five* non-collinear points, and therefore to remove five freedoms (leaving a mobility of 1). For example, as with the cylinder, the rotational freedom about a line perpendicular to the surface is now lost, because it would raise at least one of the five points off the conical surface. Moreover, the translational freedom that the cylinder allows along its axis is lost on the cone since this translation would also lift one of the five points off the surface. All five points remain in contact, and travel along parallel circles, only if the five-legged stool is rotated about the axis of the cone.

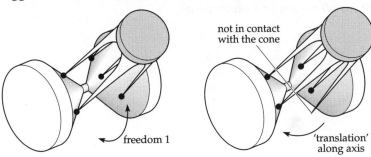

Solution 4.1

(a) This system has four links and three revolute pairs, so $n = 4$ and $j = 3$. Hence

$$M = 3(4 - 1) - (2 \times 3)$$
$$= 9 - 6$$
$$= 3$$

(b) This system has four links and three revolute pairs, so $n = 4$ and $j = 3$. Hence

$$M = 3(4 - 1) - (2 \times 3)$$
$$= 9 - 6$$
$$= 3$$

(c) This system has four links and four revolute pairs, so $n = 4$ and $j = 4$. Hence

$$M = 3(4 - 1) - (2 \times 4)$$
$$= 9 - 8$$
$$= 1$$

(d) This system has four links and four revolute pairs, so $n = 4$ and $j = 4$. Hence

$$M = 3(4 - 1) - (2 \times 4)$$
$$= 9 - 8$$
$$= 1$$

Note that the numbers given for the mobility in parts (a) to (d) are consistent with the numbers obtained in the solution to Problem 1.6.

(e) This system has four links and five revolute pairs, so $n = 4$ and $j = 5$. Hence

$$M = 3(4 - 1) - (2 \times 5)$$
$$= 9 - 10$$
$$= -1$$

(f) This system has four links and six revolute pairs, so $n = 4$ and $j = 6$. Hence

$$M = 3(4 - 1) - (2 \times 6)$$
$$= 9 - 12$$
$$= -3$$

Solution 4.2

The eight-link planar kinematic system has five binary links, two ternary links, one quaternary link and ten binary joints (revolute pairs), so $n_2 = 5$, $n_3 = 2$, $n_4 = 1$ and $g = 10$. Also $n_r = 0$ for all values of r other than $r = 2$, $r = 3$ or $r = 4$. Hence

$$M = 2g - n_2 - 3n_3 - 5n_4 - 3$$

$$= (2 \times 10) - 5 - (3 \times 2) - (5 \times 1) - 3$$

$$= 20 - 5 - 6 - 5 - 3$$

$$= 1$$

Solution 4.3

We use the direct graph representation and consider the g joints to be vertices free to move in the plane. Considered in isolation, each vertex has two freedoms, since it is located by two coordinates. Therefore, considered in isolation, the total freedom of all the vertices is $2g$. Each binary link is an edge of the direct graph and imposes one constraint on its end-points, since it prevents them from moving apart. Because we have n_2 edges, we impose n_2 constraints. Finally, we must subtract three freedoms to locate a particular edge, chosen as the fixed frame of reference. The mobility criterion is therefore

$$M = 2g - n_2 - 3$$

which is identical to that presented in Theorem 4.2 when there are only binary links present.

Solution 4.4

These systems can be obtained by using interchange graphs and listing the simple connected graphs with 4 vertices (the links) and 4 edges (the revolute pairs). There are just two such graphs — namely:

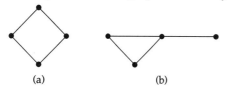

(a)   (b)

So there are just two systems:

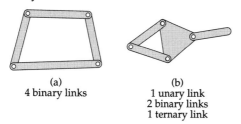

(a)
4 binary links

(b)
1 unary link
2 binary links
1 ternary link

Notice that, although both systems have mobility 1 (from Theorem 4.1), the mobility of system (b) is concentrated in the unary link and the other three links do not move relative to one another. We shall generally not deal with systems like this in which one part is a substructure that could be replaced by a single link.

Solution 4.5

The remaining eight different expansions of the two ternary joints in the diagonally braced bay are:

In all eight cases there are six binary joints, three binary links, and two ternary links, so $g = 6$, $n_2 = 3$, $n_3 = 2$, and we have

$$M = 2g - n_2 - 3n_3 - 3$$

$$= (2 \times 6) - 3 - (3 \times 2) - 3$$

$$= 12 - 3 - 6 - 3$$

$$= 0$$

as expected, since the bay is rigidly braced.

Index